Palgrave Advances in Possibility Studies

Series Editors
Vlad Glăveanu
School of Psychology
Dublin City University
Dublin, Ireland

Wendy Ross
Psychology
London Metropolitan University
London, UK

Constance de Saint Laurent
Department of Psychology
National University of Ireland, Maynooth
Maynooth, Kildare, Ireland

Palgrave Advances in Possibility Studies explores how humans engage with the possible across multiple domains including futures thinking, embodied and extended cognitive science, educational practices, creative arts, and design. This transdisciplinary series integrates different theoretical frameworks and methodological approaches to understand how possibilities are discovered, created, and enacted across different contexts and systems. Spanning future studies, creativity studies, cognitive philosophy, serendipity research, educational theory, cultural studies, and art/design practice, the series examines possibility-thinking and possibility-making as embodied, distributed processes that emerge through the interaction of minds, bodies, materials, and environments.The series welcomes proposals for short (25,000-50,000 words) and standard length monographs (80,000+ words), authored or edited.

Luis de Miranda

Crealectics as a Creative Method

A New Approach to Holistic Innovation for Compossible Worlds

palgrave
macmillan

Luis de Miranda
House of Innovation
Stockholm School of Economics Institute for Research
Stockholm, Sweden

This work was supported by funding from the European Union under the Marie Skłodowska-Curie Actions grant agreement No. 101081293SSE. It was also supported by the Turku Institute for Advanced Studies, The University of Turku and by Vattenfall R&D.

This work was supported by Vattenfall R&D Sweden and the House of Innovation/SSE Institute for Research, Stockholm School of Economics.

ISSN 3091-2571 ISSN 3091-258X (electronic)
Palgrave Advances in Possibility Studies
ISBN 978-3-032-00426-0 ISBN 978-3-032-00427-7 (eBook)
https://doi.org/10.1007/978-3-032-00427-7

This Palgrave Macmillan imprint is published by the registered company Springer Nature Switzerland AG.
The registered company address is: Gewerbestrasse 11, 6330 Cham, Switzerland

Contents

List of Figures

CHAPTER 1

Introduction: Toward Eudynamic Realities

Abstract This chapter introduces crealectics as a new creative method rooted in process philosophy and designed for holistic innovation. Drawing from the concept of the "Creal" (Creative Real), it presents a five-phase methodology—resetting, crealing, profusing, compossibilizing, and realizing—that mirrors natural creative processes. The chapter establishes the philosophical foundations, introduces the C.I.P.H.E.R. model framework (Crealectic Intelligence and Philosophical Health for Eudynamic Realities), and provides context through the Vattenfall R&D case study, positioning crealectics as an alternative to purely analytical or dialectical approaches to innovation.

Keywords Crealectics • Creative Real (Creal) • Holistic innovation • Process philosophy • C.I.P.H.E.R. model • Vattenfall R&D • Methodology • Compossibility • Innernovation • Regenerocene

We are standing at the edge of an ocean. The boundary between water and air shimmers with possibility, neither solid nor void; it is vibrating with potential. What if this liminal space of becoming is not merely a poetic metaphor but the actual texture of existence itself? Reality, at its core, isn't a collection of fixed objects but an ongoing explosive process, an unceasing creative and dynamic blast that crystallizes momentarily into the useful

L. de Miranda, *Crealectics as a Creative Method*,
Palgrave Advances in Possibility Studies,
https://doi.org/10.1007/978-3-032-00427-7_1

forms we call "things" (Bergson, 1911; Whitehead, 1978). This book, the chair or legs supporting you as you read, the thoughts forming in your mind are all temporary eddies in a vast liquid society that, as Heraclitus understood, is in constant metamorphosis (Rescher, 1996). You will never read this book in the same way twice.

If this is your first reading, the title contains both familiar words (method, innovation) and less familiar ones (holistic, compossible, crealectics). Most readers will wonder if what follows is merely another method for innovation or creative brainstorming. I would like to believe that this is not the usual methodological book—time will tell.

The following chapters are written by a philosophical practitioner, one who has spent some years in the borderlands between the triangle of abstract thought, creative authorship, and concrete action, between the dream of theory and the wakefulness of practice. Marx was wrong when he criticized philosophy for being passive: thought is not just interpreting the world; it can also be co-creating it, whether we are aware of it or not (Bohm, 1980; Varela et al., 1991). The Crealectic Method is a way to work the world out while understanding our purpose and unveiling our worldview. Hopefully, I will convince you that it is an efficient new method to embrace a plurality of lifeworlds and patiently unite them into a civilization, a common inspiring universe.

The Living Method

The crealectic method is close to the approach implicitly followed by life and evolution. When we observe closely, we may see that reality often proceeds through these motions, which also constitute the structure of the present book:

Resetting: re-originating, here and now, a point of availability for novelty;
Crealing: tapping into the creative source that supports life and possibility;
Profusing: freely exploring the exuberance of possible ways;
Compossibilizing: uniting the possibles into relational symbioses of compatibility;
Realizing: understanding and actualizing the structures that support our lifeworlds.

How cells divide and differentiate, how ecosystems balance and evolve, how stars form and perish, how cities thrive—all by following a fundamental crealectic rhythm that is here displayed as a method we can use to revive our ways of working and becoming. Whether we know it or not, we are all already, more or less, crealecticians.

When we reset, when we make tabula rasa, we prepare to focus. We delimit a zone of availability and practice that Descartes called the *cogito: I observe that I think, imagine, feel, therefore I am* (Descartes, 1637/1999). In resetting, we can momentarily forget our routines and turn inward to self-reflexivity, the intention to be mindful and the goodwill to be conscious, alive, rather than a ghost in the machine.

Thus begins the voyage of a "noonaut"—from the Ancient Greek *nous* (mind) and *nautes* (sailor). We are all noonauts, minds sailing the sea of imagination and knowledge constituting the Creative Real (Bachelard, 1958; Teilhard de Chardin, 1959). Resetting reconnects us with mind and free association as primary experiences, and we find ourselves daydreaming in a yellow submarine, exploring the sea of ultimate possibility that will be here called "Creal."

All shapes are ideas in flux, temporary agreements between mind and matter. Our embodied minds can do much more than we typically allow. They have, for generations, moved us, transformed us, transfigured the real around us. When we are in the flow, we are "crealing." The body itself is an ideational flow. We are standing in a garden in spring, watching as multicolored buds unfold, each revealing its unique pattern of dynamic harmony (Kauffman, 1995).

We are always in the Creal, even—perhaps especially—when we dream (Deleuze, 1994). This is how possibilities profuse and emerge in the creative universe—not one at a time, but in cascades of potential. And we are not just watching, we are intercreating ourselves, making choices, playing with our sense of the possible (Thompson, 2007), pruning or welcoming our potential moves, manifesting values and connecting freely what may not seem factually connected.

Amid the crealing chaos we see an explosion of alternatives, we may emulate a myriad of propositions, new ways for things to manifest and roads not taken. The universe is a possibilisation furnace (Binswanger, 1963), a profusing of forms and variations—and this creative chaos is capable of temporary harmonies, as is obvious when we look around us. In compossibilizing, we recognize that all these sparkling possibles may also form coherent, if momentary, wholes. The Creal is both the wave and the inhabited ocean in which symbiotic species abound. While it generates unlimited diversity, it also weaves connections between seemingly separate elements. Realities are not just dynamic, they can be eudynamic, in healthy balancing.

Yet the previous ocean metaphor, while evocative, may not capture the full intensity of creation. What if reality is less like flowing water in some cases and more like a constant detonation? Not a gentle stream but a repeated Big Bang, each moment exploding into the next with creative impetus? This "blastology"—the understanding of reality as explosion—offers an alternative to process philosophy's emphasis on flow. Or perhaps the reconfiguration is musical. We listen to this Music of the Spheres, a symphony where different instruments play distinct melodies, sometimes an apparent cacophony, yet together they may create a eudynamic composition, punctuated by slow or rapid metamorphoses.

The multiverse doesn't just create possibilities; it creates, here and there, compatible possibilities—possibilities that may exist together in coherent worlds. Even when actualized forms seem to somehow conflict, like fire and water or like disparate political parties; as potentials within the Creal they may coexist, sharing the same virtual origin and the same confused desire for unity. Compossibilizing is how we consciously participate in this natural tendency of the universe to create possible harmony and coherence amid diversity or cacophony.

Finally, we *realize*. This step carries a fascinating double meaning: to understand and to make real. When we truly understand something, it is almost like bringing it into reality. When we make something real, we deepen our understanding. A sculptor first sees a figure in raw marble, then gradually reveals it through careful chiseling—the figure was always potentially there, and the sculptor's art is in recognizing and releasing an idea, which is a composed cluster of possibles.

Reality isn't made of isolated objects but of dynamic relationships. When we say: "This is a cup," we're really saying: "This relates to us in a cup-like way." Everything defines itself through its relations to everything else, like words in a language that gain meaning through their differences from other words, as the structuralists have shown. The cup exists for us because it isn't a plate or a bowl; it stands out against the background of everything it is not.

This is how the universal wave function (the Creal) "collapses" into specific reality—through relationships and distinctions, local perspectives. When we realize something, we're participating in this whirlpool of relationship-making. We're helping the universe distinguish one pattern from another, allowing specific forms to emerge from the ocean of potential. A healthy world is what happens when possibilities become compatible enough to make-sense together. Each of us carries a lifeworld, shaped by the territory of our experiences and meanings. My possible, your possible, their possible—our compossible: the harmonization of our lifeworlds is an ongoing and vital process. It is complex and difficult; the crealectic method is here to help, especially, as we will see, in balancing and reuniting four essential dimensions: person, earth, work and society.

Origins and Definitions of Crealectics

The word "crealectics" might at first appear unnecessary—a neologism seeking its place in the already vast lexicon of philosophy or methodology. It resonates with "dialectics," but it takes us somewhere different. Where dialectics marches toward the synthesis of opposites like a Napoleonian warrior storming a hill (Hegel, 1807/1977), crealectics meanders like a gardener tending a fractal orchard, where each fruit contains the seed of a new world. A momentary synthesis that is gained through oppositional cognition is different than an eudynamic equilibrium sustained via compossible cognition. There are sometimes merits to dialectic thinking. In Plato's cave, dialectics leads the prisoner vertically, toward the sunny Eternal Forms above (Plato, 375 bce/1974); but in crealectics, there is no up or down, no left or right, and we recognize the shadows in the cave as part of the reality at play, no less real than the sun. In Hegel, spirit often incarnates from binary contradictions (Žižek, 2012); in crealectics, spirit thrives in diversity, asymmetric explosion and compossibility. Plato's dialectics depreciates phenomena as superficial waves (Hadot, 2002), the surface foam of abstractions; crealectics does not hierarchize the whitewater of the wave, its peak, lip, curl, trough and impact zone. We need to recognize that our three modes of thinking and world-making are equally important: analytic, dialectic, crealectic. We divide our territories into known parts and infer future realities from past data (analytics). We negotiate our understandings through synthetic tensions between thesis, antithesis, and via agonistic debate (dialectics). We also create and are created at the same time in a process of connecting and disconnecting the possibles incessantly—and this crealectic process is not deterministic because it makes leaps between disparate worlds. Analyticians are world-dissectors, ready to sew anew what they have divided. Dialecticians are world-evaluators, playing billard with the possibles until a new game is afoot. Crealecticians are explorers leaping from world to world while holding the balance, and in thus doing, manifesting new worlds. Dissecting, evaluating and inter-creating: these three activities when harmonized are the cognitive engine of holistic innovation.

This "book" is a living experience, a momentary contraction of the Creal into functional shape, printed paper or series of digital ciphers. It is not only a thing but a process serving your ideas and decisions, a node in a vast network of transformations. Where analytic thinking wields a scalpel—cutting situations into known bits—and dialectics hypnotizes us with a binary pendulum of theses and antitheses, crealectics offers an ecosystem weaving threads of potential into the beehive of the actual, and vice versa, in a movement of zig zag (Deleuze, 1994), or an infinite feedback loop.

Crealectics emerges from three etymological roots: *creation*, *real*, and, alternatively, depending on the philosophical tonality of the situation at stake, *logos* (logic) or *ektos* (outside). It stands alongside analytics and dialectics. Where analytics divides and dialectics opposes, crealectics flows and connects by intercreating. It engages directly with reality's creative pulse, embracing both its essential unity (the diverse Manifold, logically, is One manifold) and its infinitely varied expressions. Cosmologically, there are many worlds, some adjacent, and some invisible to each other. Outer worlds of interactions and inner lifeworlds of imagination and feeling. Crealectics connects and transcends these orders to let novelty emerge pragmatically and meaningfully.

The "Creal" refers to the continuously Creative Real, deep reality understood as process philosophers do: not as stable objects but as a flux of continuous creation and actualizing negotiations out of the ultimate possible. As quantum physics revealed with its universal wave function and probability fields, beneath the apparent solidity of matter lies a sea of potential waiting to be actualized, and actualized worlds retain within themselves the capacity to transform or mutate anew. The Creal is both the source of this potential and the whole of the universe, a unified field of becoming, structuring and destructuring.

What furthermore distinguishes crealectics from usual analytic thinking is its understanding that humans don't observe this process from outside; we participate in it. Every thought or acquiescence, every action or reaction, every invention or submission is our conscious or unconscious contribution in the unfolding of reality. Crealectic intelligence is the deliberate and proactive practice of this participation, organizing diverse and harmonious realities from an embodied sense of possibility and dynamic compatibility.

The crealectic approach centers on developing authorship in what we do, being deeply active, alive, perceptive of difference rather than merely productive of sameness along the routines of our systems. Creativity isn't primarily about tools or artifacts but about self-innovation, or what we will call in what follows "inner-novation": enhancing our capacity to tap into the creative flow of the multiverse via the phenomenology of our gratitude for being alive and embodied in our "proprioconception," the conception that we become who we are.

Central to this framework is what I call "authoricity," a concept integrating authenticity, authorship, and authority (see Chap. 6) (Fig. 1.1).

Authenticity refers to being true to oneself and one's values; authorship involves creating with originality and purpose; authority encompasses the ethical influence one exercises in the world. When these three dimensions align, we experience a state of potent creative agency that transcends mere technical innovation. Before we dive anew into this matrix in chapter 6, try to evaluate for yourself your level of authoricity as a person and in your work: how self-honest are you and can you keep your integrity of

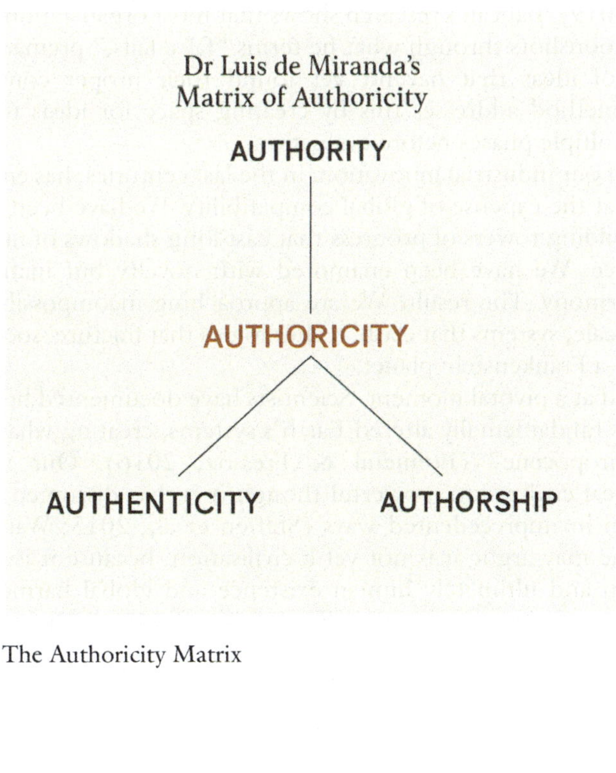

Fig. 1.1 The Authoricity Matrix

character across situations (authenticity)? Do people listen to you and do you have the courage to manifest your higher purpose (authority)? To what extent would you say your life is your oeuvre, rather than being an existential puppet (authorship)?

Unlike conventional approaches to innovation or creativity that focus primarily on external products, measurable outcomes, or technological development, crealectics emphasizes the transformation of consciousness itself as the necessary foundation for meaningful external change. To relate with the new differently and holistically, we must first perceive and feel differently, and this experiential shift begins with recognizing our participatory relationship with reality's sense-making and creative process. If you think that adaptation is the only way of being, this book is (not) for you.

The Need for Holistic Innovation

Innovation researcher Safi Bahcall's analysis of breakthrough innovations reveals a crucial distinction between what he calls "loonshots," widely dismissed ideas that change the world, and incremental improvements (Bahcall, 2019). Bahcall's research shows that most organizations inadvertently kill loonshots through what he terms "false fails," premature abandonment of ideas that haven't yet found their proper context. The crealectic method addresses this by creating space for ideas to develop through multiple phases before evaluation.

Much of our industrial innovation, in the last centuries, has emphasized possibility at the expense of global compatibility. We have been architects of silos, building towers of progress that cast long shadows of unintended consequence. We have been enamored with novelty but inattentive to general harmony. The result? We are approaching incompossibility on a planetary scale; systems that clash, technologies that fracture, societies that fragment—a Frankenstein planet.

We stand at a pivotal moment. Scientists have documented how human activity has fundamentally altered Earth's systems, creating what they call the "Anthropocene" (Bonneuil & Fressoz, 2016). Our industrial-technological civilization, powerful though it is, has disrupted planetary equilibrium in unprecedented ways (Steffen et al., 2015; Waters et al., 2016). One may argue it is not yet a civilisation, because it is hostile to non-human and ultimately human existence and global harmonization. The Frankenstein monster of globalization is searching for a soul.

The innovative mindset that built our current world, focused on making things possible without considering whether they're compatible with other existing or future possibilities, is indeed proving dangerously insufficient. Local technological marvels may evolve into global catastrophic risks. Short-term "brilliant" solutions create long-term less shiny problems. We will not fix the "anthrobot" with another "anthrobot," we will not solve human superficial intelligence with artificial intelligence. The fragmented technological mindset that created our ecological crises cannot, through its own internal logic, resolve them.

What we need is a transition toward the "Regenerocene," an era where human and non-human life can flourish through mutual regeneration (which may include death as a moment of the process) and intercreation. This requires more than technological fixes. It demands indeed, not only innovation but "innernovation," the exploration and transformation of the ideas that shape our personal relationship with reality itself, and us within it. Educational theorist Ken Robinson argues that our current systems systematically educate people out of their creative capacities, creating what he calls an "academic inflation" that privileges certain forms of intelligence while neglecting others essential for innovation (Robinson, 2011). Similarly, philosopher Anthony Weston suggests that much of our innovation crisis stems from poorly framed problems rather than inadequate solutions, advocating for what he calls "problem-crafting" as a foundational skill (Weston, 1992). These insights align with the crealectic recognition that genuine innovation requires not just new answers but fundamentally new ways of questioning and exploring.

This inner work isn't separate from practical innovation but is its necessary foundation. As the controversial industrialist Henry Ford (1929) wrote: "Perhaps the most one can hope for now is to drive home a conviction that as a people we have not done much thinking. If we think we have been thinking, and then find out that we have not, the jolt of discovery may be of service to us. The secrets of life are opened to the thinker." And thinking is necessary but not sufficient; one must think holistically by replacing our so-called problems in their intercreative context with persons, the earth, our ways of working and our societies.

What we need most urgently is a compossible mindset: a way of thinking that fosters diverse possibilities while ensuring they harmonize dynamically with one another. The seventeenth century mathematician and philosopher Gottfried Wilhelm Leibniz, who introduced compossibility into the history of ideas (1989), inspires us to think of "the best of all possible worlds" not as perfection but as holistic optimization—maximizing diverse goods in intertwined becoming. A compossible mindset seeks this dynamic harmony not by systematically eliminating differences but by discovering how contradictions might (or not) be transcended together in a richer unity, a oneness that preserves diversity.

Many of us are worried by the cacophony of our globalized chaos. But it's not too late to intercreate our shared planetary civilization. The regenerative mindset connects to what positive psychology research identifies as "post-traumatic growth," the phenomenon where healing and development become interdependent (Fredrickson, 2013; Seligman, 2011). Like forests that thrive after fire, or coral reefs that adapt to warming seas, resilience sometimes emerges not despite chaos, but through it. Studies of communities recovering from collective trauma show enhanced resilience and social cohesion through collaborative reconstruction efforts (Osofsky et al., 2018). Similarly, restoration ecology demonstrates that properly directed human activity can facilitate ecosystem healing rather than degradation (Perring et al., 2015).

United scientific and philosophical perspectives suggest humanity must develop a shared mindedness, a collective consciousness beyond the fragmented values and competitive practices characteristic of the Anthropocene (Steffen et al., 2011). Others, grounded in developmental psychology, contend that such universal transformation cannot happen instantaneously, if it's possible at all (Kegan, 1982). What remains clear is that the still mainstream innovative paradigm, one that privileges possibility without compossibility, is increasingly maladaptive in an interconnected world.

The C.I.P.H.E.R. Model

I have been developing a crealectic approach to philosophy since 2008 in both non-fictional and fictional texts. All started with the discovery of the Creal notion in my science-fiction novel, *Paridaiza*, first written in French and later translated and published in the USA (de Miranda, 2020). Speculative fiction, like dreams, is a laboratory for wild ideas that might not emerge out of analytic contexts of seriousness, routines and deadlines.

The Crealectic Method is one of two interconnected approaches within the C.I.P.H.E.R. model I've developed over the past years. CIPHER is both compass and crucible; a tool to navigate uncertainty and a vessel to alchemize ideas into action. CIPHER is an acronym that stands for

Crealectic Intelligence and Philosophical Health for Eudynamic Realities. It is a framework for both personal and organizational transformation.

CIPHER proposes that crealectic intelligence works in dynamic relationship with philosophical health, the flourishing coherence between our meanings, values, and actions. Together, they generate "eudynamic realities," experiences that enhance rather than contract our senses of the possible and the compossible, at the crossroad of personhood, earthly becoming, our ways of working to transform the world, and our societies. I won't expand here on the SMILE_PH method (Sense-Making Interviews Looking at Elements of Philosophical Health) as I published another book about it: *Philosophical Health: a Practical Introduction* (de Miranda, 2024). In what follows, we will indeed focus on the second method of the CIPHER model (Fig. 1.2), the Crealectic Method.

Unlike conventional innovation approaches focused on external prod ucts, CIPHER emphasizes inner transformation as the foundation for meaningful external change. The framework integrates insights from process philosophy, cognitive science, and creative ecology to provide practical methods for engaging more consciously in the co-creative relationship with reality. This integration of philosophical depth with practical action reflects what business philosopher Peter Koestenbaum calls "philosophic consulting," the application of fundamental philosophical insights to concrete organizational challenges (Koestenbaum, 1991). Philosophical thinking can transform not just what leaders do but how they understand their role in creating organizational and productive reality.

The philosophical architecture of CIPHER echoes with diverse intellectual traditions—process philosophy (Prigogine, 1997; Whitehead, 1978), existential psychology (Binswanger, in Needleman, 1963), neuroplasticity research (Doidge, 2007), and creative ecology (Capra & Luisi, 2014; Harris, 2000). These perspectives converge in recognizing that consciousness is not separate from reality's creative processes but actively participates in the unfolding of our more or less rich realities.

Neuroscientific research shows that contemplative practices can enhance cognitive flexibility and attention regulation (Lazar et al., 2005; Siegel, 2012). Through deliberate cognitive development, we can engage more consciously in our intercreative relationship with reality, transforming both ourselves and part of our lifeworlds in the process. Our brains are not static organs but plastic collaborators, reshaping themselves with every persistant thought, perhaps not like clay responding to the potter's hand but with real long-term plasticity.

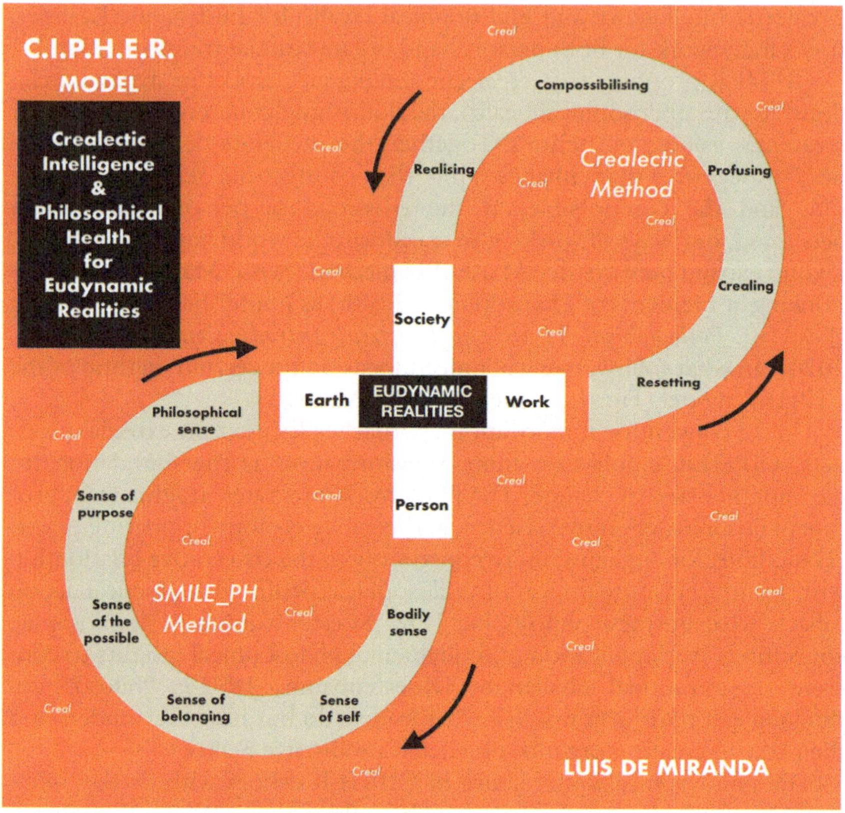

Fig. 1.2 The C.I.P.H.E.R. model framework

At the center of the CIPHER model is the ideal of enriching our realities by deploying holistic actions on four axes: person (who we are individually, that is embodied), earth (where we are collectively, that is extended), work (how we transform the world, in enaction), and society (how we live together, that is embedded). The crealectic method keeps an eye on these dimensions by design, such that the solutions it may help unfold are not only possible but also compossible. Not only is our cognition *embodied* in personhood, *extended* within earthly context, *enacted* in work understood as world transformation, and *embedded* in societies (as per the 4E cognition perspective), but also these 4Es (Embodied, Extended, Enacted and Embedded) must be integrated to generate a good dynamic balance, or *eudynamia*.

The Vattenfall R&D Case Study

Throughout this book, Vattenfall, one of Europe's largest and greenest energy companies, serves as both case study and metaphor. As a Sweden-based international organization transforming diverse energy sources into usable power, it mirrors metaphorically the fundamental process at the heart of crealectics: transforming creative potential (Creal) into realized actuality (Real). The energy sector's transformation—or lack thereof—mirrors what systems leader Giles Hutchins calls "nature-inspired leadership," moving where possible from mechanistic, extractive models to regenerative approaches that align with natural patterns (Hutchins, 2022). I believe that organizations like Vattenfall represent laboratories for crealectic thinking.

In autumn 2019, I began working with the Strategy and soon after the Research and Development (R&D) unit of Vattenfall, first conducting philosophical health conversations with some of their leaders. Via their R&D division, what began as individual philosophical coaching evolved into a network for long-term thinking and strategic action. The timing proved significant: Vattenfall was navigating a profound transition as former fossil fuel giants entered their renewable energy market and societal demand for clean electricity surged.

Vattenfall's power portfolio spans wind, hydroelectric, solar, nuclear, and biomass technologies. Their vision centers on enabling "fossil-freedom" or "fossil-free living." Their transition is still work in progress, and their progress sometimes erratic, uncertain, and slowed by external and internal obstacles, but it could mirror the crealectic journey: from fossilized thinking to fluid innovation, from extraction to intercreation. This manifests in practical innovations like their contribution to initiatives toward fossil-free steel production processes that demonstrates how industrial decarbonization can emerge through creative reimagining of established protocols.

Encouraged by a few visionary leaders like Andreas Regnell, Jonas Wilde, Holger Ecke and Ambra Sannino, I developed and tested the SMILE_PH method and the crealectic method in several individual sessions or collective workshops at Vattenfall R&D, since 2023. These applications explored ambitious territories: examining for instance "regenerative consciousness" or investigating "values underlying purposeful growth." This practical form of research exemplifies what philosophers call "field philosophy" or "action philosophy," philosophical practice applied directly to real-world challenges.

The energy sector provides an ideal testing ground for creative methods because it grapples with multidimensional challenges spanning technology, environment, and society. The domain faces particularly complex challenges at the intersection of technological innovation, environmental sustainability, and societal transformation—what complexity theorists call "wicked problems" that resist simple solutions (Miller & Page, 2007). These are Gordian knots that defy the sword of linear logic but may yield to the patience of systemic creative weaving. These multidimensional challenges will benefit from integrative methodological frameworks that can embrace both technical precision and holistic awareness. The crealectic method offers such a framework, providing a structured yet flexible approach to innovation that acknowledges the intricate interdependencies of our complex world, as well as the importance of human agency and consciousness. Person, Earth, Work and Society are so intertwined today that considering them in isolation is no longer a viable option.

My action-research interventions at Vattenfall are gradually evolving into a collaborative network oriented toward facilitating long-term thinking and strategic action for holistic energy systems. As I am writing this book, it is uncertain if I will succeed in developing the crealectic mindset with more than a group of persons or projects. Circumstances may create an optimal environment for methodological innovation, allowing the crealectic approach to demonstrate its practical efficacy in a real-world context. But like other field philosophers before me, I might be too idealistic in my ambition to change the world. Change demands extreme patience against the resistance of old ways of doing and the mechanical protocols of (un)thinking. But which healthy transformation does not take time?

Vattenfall is an early adopter, a pioneer in crealectics. Other corporations might soon take over. It is too soon to tell. It's up to us, and this includes the readers of this book.

Overview of the Chapters

The crealectic method offers a distinctive approach to thinking, planning, and acting that fosters both diversity and eudynamic harmony. It unfolds through five interconnected phases that mirror the ancient wisdom of nature itself, perhaps the very steps, one could argue, through which life evolves, adapts, and flourishes:

1. Resetting (Chap. 2): Clearing the mental slate to create space for fresh perception and approach *tabula rasa*. This is the cosmic pause, the winter of the mind when old forms dissolve back into a vibrant blank slate.

Resetting involves suspending our habitual patterns of thought, setting aside what cognitive scientists call "autopilot cognition" (Kahneman, 2011). To reset is the peaceful pause between breaths, the hush before dawn, where the old-world dissolves and the new one hesitates on the threshold. This doesn't mean achieving a perfect blank slate, probably impossible for embodied beings with complex histories, but creating enough mental space, cognitive rest and inner availability for something new to emerge. Neuroscience shows that contemplative practices—such as meditation or reflective activities like reading or writing poetry—can activate the brain's default mode network during rest states and that creative thinking involves dynamic cooperation between routinely opposing neural networks (Beaty et al., 2018; Raichle et al., 2001).

2. Crealing (Chap. 3): Feeling and reconnecting with the Creal, the creative flow that underlies reality. This is the embryonic state of becoming, the spring of both consciousness and life when we sense the stirring of ultimate and infinite possibility.

Crealing invites us to feel rather than merely think about the creative flow that constitutes reality. This is not abstract theorizing but embodied knowing and proprioceptive feeling, the way a dreamer intuits possibilities in flux, or a diver engages within the wave's intention. It's a felt recognition of life as an intercreative network of relationships where humans participate rather than merely observe. Abundance as primeval. This reconnection amplifies our sense of what's possible, a feeling of excitation and gratitude associated with enhanced well-being and cognitive capacity (Fredrickson, 2013; McGonigal, 2015).

3. Profusing (Chap. 4): Letting ideas and possibilities pour forth without censorship. This is summer's exuberance, when life expresses itself in riotous diversity and unapologetic generative abundance of forms and behavior.

Profusing encourages the uncensored expression of ideas and possibilities, similar to the Surrealists' techniques for accessing the generative potential of the unconscious. Here, the mind becomes a miniature Big Bang, scattering stardust of ideas without yet pausing to shape or discern the constellations. Research on creative cognition suggests that divergent thinking processes engage distributed neural networks that are associated with novel combinations (Jung et al., 2013). This phase prioritizes abundance over immediate practicality.

4. Compossibilizing (Chap. 5): Discovering how diverse possibilities can coexist and enhance each other when configured into a meaningful

network. This is autumn's elegant selection, when what matters is preserved and what doesn't is gracefully released in optimizing harmony.

Compossibilizing draws on Leibniz's concept that a coherent world comprises only elements that can coexist without fundamental contradiction (they are possible together). This is the alchemy of dynamic harmony, where seemingly discordant notes find their evolving place in the symphony, where the jagged edges of innovation are smoothed into a mosaic of mutual pruning. This phase identifies potential synergies among diverse ideas, looking for patterns of compatibility and mutual enhancement. It's similar to how complex adaptive systems generate emergent properties through the interaction of their elements (Holland, 2014; Kauffman, 2019).

5. Realizing (Chap. 6): Both comprehending and implementing, understanding and manifesting what has emerged. This is the harvest of all the seasons at once, when potential becomes actual and the cycle prepares to begin anew on a higher level of the spiral of eudynamia.

Realizing embodies a dual meaning: understanding what was previously unclear and bringing potential into actuality. This is where the abstract becomes anchored in realities, where the architect's sketch prepares to meet the weight of stone, regenerative brick, and fossil-free steel. It integrates reflection with implementation, acknowledging that effective manifestation requires both conceptual clarity and practical strategy (Dewey, 2005; Schön, 1983). Realization is the moment when the invisible may be unveiled as a new local manifestation of the generative dynamic relationship between person, earth, work and society.

These steps may seem sequential, and they are conducted so in workshops or individual dialogue. But in reality, they may interweave and overlap due to the complex, non-linear nature of creative processes. They form not a staircase but a spiral; we return to each phase with deeper insight, like seasons revisiting a landscape, each time revealing new contours of the infinite feedback loop.

Throughout the following chapters, we'll explore each step in detail, examining philosophical or scientific foundations as well as practical applications. The crealectic method is not merely a technical procedure but a holistic approach to innovation that recognizes the profound interconnection between consciousness, creativity, and reality itself. By engaging daringly with this integrated methodology, individuals and organizations can develop approaches to innovation that honor creative potential while fostering harmonious relationships among beings and becoming. The methodology is performative and exploratory, revealing news ways of doing and thinking about the world as it unfolds and expands. It is my intention that this book may serve as a manual, but I have also developed a thorough certification in the Crealectic Method (as well as in the SMILE_PH method) by which I train practitioners in face-to-face online sessions of in-depth mentoring.

Whether we are an engineer, a decision-maker, or a student, still undecided about who we are and what we ought to do, the CIPHER model and this book will train us to see reality differently, not as a collection of fixed objects to manipulate, but as a creative process in which we already participate. We are not a bystander to creation but a co-conspirator, whispering compossibilities into the ear of the universe, or vice versa. By becoming more conscious of this participation and intercreation, we can contribute more effectively to creating what Leibniz called "the best of all possible worlds"—not a perfect utopia, but a eudynamic composition of the richest possible diversity, unified into a shared civilization.

In contrast to many innovation methodologies that remain abstract or purely theoretical, the crealectic method has been tested and refined through practical application in organizational contexts, at Vattenfall and other companies or institutions. In particular, the concrete early experiences gained through testing and implementing this approach with Vattenfall's R&D team have informed both its theoretical foundations and its practical protocols. The result is a methodology that bridges philosophical depth with pragmatic utility, but without fetishizing the production of results. If it can be said that engineers *mind the gaps* in reality, then attention should be given to the mind itself, not just the gaps. Ours is a world where the central question shifts from What can I extract? to What can we co-create and to serve what higher purpose?

As we progress through each chapter, we'll find not only conceptual explanations but also, here and there, specific exercises and practical guidelines for experimenting with each phase of the method. Whether you're an organizational leader seeking new approaches to innovation, a researcher exploring creative methodologies, or an individual interested in expanding your own creative capacities, the crealectic method offers a structured yet flexible framework for engaging more consciously with reality's enriching and eudynamic potential.

In fact, the journey we're embarking on transcends conventional categories of "practical" versus "theoretical." It invites us to recognize that our deepest philosophical insights can transform our most concrete actions, and our everyday practices should inform our most profound understandings. In this entanglement between thought and action, between possibility and actuality, we discover new paths toward creating not simply innovative products or services, but also more harmonious ways of being in the world, together among earthlings.

We are all, intuitively, crealecticians participating in the unfolding of the world. What this method offers is not a foreign doctrine, but a conscious rendering of something native, a bringing-to-light of the tacit creativity already at work in our lives, even if sometimes hindered by blockages. The crealectic method is not something imposed upon reality but rather a conscious recognition of how reality itself unfolds if we look and listen carefully. By aligning our innovative practices with the fundamental rhythms of creation, we may discover not just new products or services, but new forms of being alive.

As we stand at the edge of an ocean, I invite you to reflect actively upon the wholeness of your lifeworld. Prepare to make it your oeuvre. What if we could become the first global and compossible civilization that ever existed? What if the earth could become what it is: a cosmic masterpiece?

References

Bachelard, G. (1958). *The poetics of space*. Beacon Press.

Bahcall, S. (2019). *Loonshots: How to nurture the crazy ideas that win wars, cure diseases, and transform industries*. St. Martin's Press.

Beaty, R. E., et al. (2018). Robust prediction of individual creative ability from brain functional connectivity. *Proceedings of the National Academy of Sciences, 115*(5), 1087–1092.

Bergson, H. (1911). *Creative evolution*. Henry Holt and Company.

Binswanger, L. (1963). *Being-in-the-world: Selected papers of Ludwig Binswanger*. Basic Books.

Bohm, D. (1980). *Wholeness and the implicate order*. Routledge.

Bonneuil, C., & Fressoz, J. B. (2016). *The shock of the Anthropocene: The earth, history and us*. Verso Books.

Capra, F., & Luisi, P. L. (2014). *The systems view of life: A unifying vision*. Cambridge University Press.

de Miranda, L. (2020). *Paridaiza*. Snuggly Books.

de Miranda, L. (2024). *Philosophical health: A practical introduction*. Bloomsbury Publishing.

Deleuze, G. (1994). *Difference and repetition*. Columbia University Press.

Descartes, R. (1637/1999). *Discourse on method and meditations on first philosophy*. Hackett.

Dewey, J. (2005). *Art as experience*. Perigee Books.

Doidge, N. (2007). *The brain that changes itself: Stories of personal triumph from the frontiers of brain science*. Viking.

Ford, H. (1929). *My philosophy of industry.* Coward-McCann.
Fredrickson, B. L. (2013). Positive emotions broaden and build. *Advances in Experimental Social Psychology, 47*, 1–53.
Hadot, P. (2002). *What is ancient philosophy?* Harvard University Press.
Harris, P. (2000). *The work of the imagination.* Wiley-Blackwell.
Hegel, G. W. F. (1807/1977). *Phenomenology of spirit.* Oxford University Press.
Holland, J. H. (2014). *Complexity: A very short introduction.* Oxford University Press.
Hutchins, G. (2022). *Leading by nature: The process of becoming a regenerative leader.* Wordzworth Publishing.
Jung, R. E., et al. (2013). The structure of creative cognition in the human brain. *Frontiers in Human Neuroscience, 7*, 330.
Kahneman, D. (2011). *Thinking, fast and slow.* Farrar, Straus and Giroux.
Kauffman, S. (1995). *At home in the universe: The search for the laws of self-organization and complexity.* Oxford University Press.
Kauffman, S. (2019). *A world beyond physics: The emergence and evolution of life.* Oxford University Press.
Kegan, R. (1982). *The evolving self: Problem and process in human development.* Harvard University Press.
Koestenbaum, P. (1991). *The philosophic consultant: Revolutionizing organizations with ideas.* Jossey-Bass.
Lazar, S. W., et al. (2005). Meditation experience is associated with increased cortical thickness. *Neuroreport, 16*(17), 1893–1897.
Leibniz, G. W. (1989). *Philosophical essays.* Hackett Publishing.
McGonigal, K. (2015). *The upside of stress: Why stress is good for you, and how to get good at it.* Avery.
Miller, J. H., & Page, S. E. (2007). *Complex adaptive systems: An introduction to computational models of social life.* Princeton University Press.
Needleman, J. (1963). *Being-in-the-world: Selected papers of Ludwig Binswanger.* Basic Books.
Osofsky, J. D., et al. (2018). Building resilience after disasters through the youth leadership program: The importance of community and academic partnerships on youth outcomes. *Progress in Community Health Partnerships: Research, Education, and Action, 16*(1), 65–72.
Perring, M. P., et al. (2015). Advances in restoration ecology: Rising to the challenges of the coming decades. *Ecosphere, 6*(8), 1–25.
Plato. (375 BCE/1974). *Republic.* Hackett.
Prigogine, I. (1997). *The end of certainty.* Free Press.
Raichle, M. E., et al. (2001). A default mode of brain function. *Proceedings of the National Academy of Sciences, 98*(2), 676–682.
Rescher, N. (1996). *Process metaphysics: An introduction to process philosophy.* SUNY Press.

Robinson, K. (2011). *Out of our minds: Learning to be creative* (Revised ed.). Capstone.
Schön, D. A. (1983). *The reflective practitioner: How professionals think in action.* Basic Books.
Seligman, M. E. P. (2011). *Flourish: A visionary new understanding of happiness and well-being.* Free Press.
Siegel, D. J. (2012). *The developing mind: How relationships and the brain interact to shape who we are.* Guilford Press.
Steffen, W., et al. (2011). The Anthropocene: Conceptual and historical perspectives. *Philosophical Transactions of the Royal Society A, 369*(1938), 842–867.
Steffen, W., et al. (2015). The trajectory of the Anthropocene: The great acceleration. *The Anthropocene Review, 2*(1), 81–98.
Teilhard de Chardin, P. (1959). *The phenomenon of man.* Harper & Row.
Thompson, E. (2007). *Mind in life: Biology, phenomenology, and the sciences of mind.* Harvard University Press.
Varela, F. J., Thompson, E., & Rosch, E. (1991). *The embodied mind: Cognitive science and human experience.* MIT Press.
Waters, C. N., et al. (2016). The Anthropocene is functionally and stratigraphically distinct from the Holocene. *Science, 351*(6269), aad2622.
Weston, A. (1992). *Toward better problems: New perspectives on abortion, animal rights, the environment, and justice.* Temple University Press.
Whitehead, A. N. (1978). *Process and reality.* Free Press.
Žižek, S. (2012). *Less than nothing: Hegel and the shadow of dialectical materialism.* Verso.

CHAPTER 2

Resetting

Abstract Resetting involves the deliberate clearing of mental space to create availability for novelty and fresh perspectives. This chapter explores how cognitive and emotional reset practices—ranging from Cartesian doubt to poetic immersion—can interrupt habitual patterns and autopilot thinking. Drawing on philosophy of mind, cognitive science and research on neuroplasticity and contemplative practices, it demonstrates how resetting creates the foundational conditions necessary for genuine creative engagement, offering ten practical exercises for individual and organizational application.

Keywords Resetting • Tabula rasa • Mental reset • Cartesian doubt • Neuroplasticity • Autopilot thinking • Mindfulness • Psychological safety • Cognitive flexibility • Contemplative practice

L. de Miranda, *Crealectics as a Creative Method*,
Palgrave Advances in Possibility Studies,
https://doi.org/10.1007/978-3-032-00427-7_2

Fig. 2.1 The resetting phase symbol

We are standing at the edge of a precipice. Behind us lie the familiar territories of habit, routine, and comfortable thought. Before us stretches the vast unknown of possibility or nothingness. The first step of the crealectic method invites us to do something perhaps counterintuitive for those who are eager to produce solutions: to pause and reset our minds, to clear the accumulated debris of perception, to create a tabula rasa from which genuine innovation might emerge (Fig. 2.1).

In our fast-paced global world, our minds operate in a relentless cycle of processing disparate information, making repeated decisions, and grappling with an array of emotions that are sometimes due to an unprecedented level of cognitive noise (Levitin, 2014). This mental whirlwind sometimes culminates in overload or apathy, decision paralysis or reactive anxiety. We sometimes become like Nietzsche's camel, carrying too much information or memories on our shoulders. Resetting offers a radical alternative, a conscious and momentary ***deworlding*** that creates availability for novelty and a fresh start.

The Freedom of Mental Reset

Our mind tends to meander from worry to rumination, a babble that keeps us superficially focused, considering both the past and future, searching for patterns like meerkats on the lookout, blind-spotting bad habits. Can we ever truly reset? There are moments of grace, like reading a poem that resonate deep within us; we might be sitting on a sunny bench and looking at a lake, talking with a dolphin in our imagination, or simply breathing the force of life, feeling the simplicity of being without having to prove anything for an instant.

What is primarily reset in crealectic collective workshops or individual sessions is, ideally, the default utilitarian mindset of the participant. We should avoid as much as possible that dull routines or defeatist prejudices are brought into the process of innovation. Even computers or phones occasionally needs a reboot to function at their best—our minds even more so. We benefit immensely from a purposeful reset into a state of availability rather than autopilot pursuit (Newport, 2016). A mental reset isn't just about taking a short coffee break or setting off on a walk. It's also a deliberate effort to cleanse one's mind, refocus on the present, life, peace, or nothingness. It's about renewing one's intentions for upcoming genuine exploration and discovery, by silencing burdened connections.

We tend to consider our lifeworld as if it were the only and universal reality. Resetting is like noise cancellation from our current social world, be it the world of the company we work for or the world of our nation, family, social media platform or profession. It's a liberation from the desperate impulse for control or performance, or from the prison of habit we've sometimes built around ourselves. It's about breaking out of the "anthrobot" that has been programmed in and around us by society, culture, natural instinct or personal history. But it's not only freedom *from*—it's also freedom *to*: one cleans the slate (*tabula rasa*) to open the space for a freedom to imagine, to experience a rebirth or the connection to something primordial, a spacetime aside from fear or confusion. In our world of digitally bombarded attention, we may suffer from mental obesity, an excess of cognitive stimuli that burdens our neuronal connections and prevents the kind of fecund minimalism that facilitates clarity.

Our worlds are formatted in one way or another, but never completely closed to the spiral of an escape, just because life circulates in our veins. Life itself is constantly resetting via birth and death. In nature, a great reboot is continuously ongoing. Life makes the infinite and the pluripotent reappear constantly at the corner of a field, for instance combining the roots that are dormant. We must gratefully welcome the sense of freedom that runs underneath our stress. We were once newborns, and in a sense, every morning is a rebirth. What is a birth if not an explosion of possibility?

This need for deliberate mental reset aligns with what productivity researchers Jake Knapp and John Zeratsky call "time shifting," the conscious redesign of our relationship with time and attention to create a focused, intentional engagement with what matters most (Knapp & Zeratsky, 2018). Their research shows that our default relationship with technology and information often prevents the kind of mental availability that creative work requires.

Ideally, a reset bestows both a letting-go and a vigilance, allowing us to view situations somewhat beyond the cloud of necessity, bias, habit, or partisanship. Our willpower is not an unlimited battery. With a refreshed mind, we prepare ourselves to tackle our inner creativity with newfound desire and optimism (Baumeister et al., 1998). On an emotional plane, resetting is akin to an inner balm mitigating the effects of stress and anxiety, leading to a balanced emotional state. Furthermore, daily resetting engenders a positive ripple effect on our overall well-being, from improved physical health to reduced stress (Hunter & Wu, 2016). A trained person can reset amid confusion, taking the subway at rush hour and observing the attitudes around, or repeating a personal motto to pause the meandering of vain attention.

At the end of this chapter, I will propose ten exercises for resetting, but the list is infinite and should be personalized by self-exploration or in dialogue with a trained crealectic practitioner. We are not Buddhas—resetting will never be perfect; yet to know that nothing is ever perfect is already resetting. Numerous studies have shown that practices like short but daily moments of meditation, nature immersion, or uncompetitive art engagement can lead to enhanced neural connections, reduced cortisol levels, and improved mood (Berman et al., 2008; Luders et al., 2009; Stuckey & Nobel, 2010). Research in neuroplasticity confirms that intentional mental practices can partly reshape our brain structure (Davidson & Begley, 2012). By aiming for an asymptotic blank slate, we're allowing our brains to relativize habitual patterns, paving the way for a state of mindful innernovation.

Resetting as Reconnection

What is reset when we reset? Perhaps it is not the mind itself but our relationship to it. We reawaken to the fact that we are observers of our own thoughts, not merely puppets moved by their strings. This reawakening is the core of consciousness, the ability to step back and watch our mental processes unfold, to choose which to engage with and which to let pass by.

One of the most efficient ways to achieve a philosophically healthy resetting is the one proposed by seventeenth century philosopher René Descartes in his *Meditations* and the *Discourse on Method* (1999) It is known under the name of cogito. Cogito means in Latin, *I think.* "I think, therefore I am": with this simple formula, Descartes performed a sort of gigantic resetting in just a few days, with quill pen, ink, paper, and a daring imagination. First, he considered that all the available knowledge of his time, constituted by Aristotelian metaphysics, Christology, geometry, as well as the emerging sciences of physics, optics, biology, might be an illusion, and that even mathematical truths could be the effect of a cosmic or evil hypnosis.

Let's pretend, Descartes suggests, that we can doubt everything: we can look at the world around us and decide it is an illusion. Perhaps the people passing by in the street are automata, robots who at a distance may appear human, whatever is meant by human. Perhaps everything is a dream. We can certainly display a radical doubt, as the skeptics do. However, one thing that the thinker cannot doubt is thought itself as it is performed in the mind. If I wonder if I am dreaming or perceiving reality, this questioning itself is absolutely undoubtable. If I doubt that I am doubting, I am still exercising my cognition. When we reset, we reconnect with the solid ground of our personal consciousness prior to any dogma or vague belief, the capacity to freely observe and experience the world out of the habits, ideologies, and routines that have shaped our biography from the outside. Even if I doubt that I can think or reset, this itself is a form of first-person mentation.

The cogito is a meditative re-awakening that can be achieved instantaneously, like a bird in our head that we are constantly catching and letting go. Contrary to what was said by anti-intellectualists, thinking is not fully abstract: it is a reconnection with our meaningful presence in the world. *Vivo, ergo sum.* I am, I live, I wonder and can wander as well as cogitate. This attentiveness to our perspective upon the world is the most philosophical form of resetting. It's not about thinking this or that but simply realizing that we are considering the world in the first person, as a sense-maker. Resetting is always subjective, and it reconnects us with our active subjectivity in a world in which we are too often made to be or feel like objects, like the automata of our everyday—real and false—statistically correct obligations.

Our personal singularity, what makes us who we are as experiential and unique subjects, is sometimes forgotten as we seek to adapt to society's needs and demands. But the more time we spend on earth, the more unique we are. Every new day adds new personal experiences to our cumulated processing of reality, and even if many of these experiences are common to most humans and even most animals—watching a blue sky or avoiding to look at the sun, for instance—other experiences are more specific. And even if they were not so singular, the order and assembly of our experiences is ultimately unique, thus creating an experiential pattern we may call our own, in a spacetime no one else occupies exactly. Resetting is not an ego-trip, but rather a remembering that the person that we are is what remains after we eliminate all that makes us measurable.

It is rather surprising how indifferent we are in general to the singularity of our self, beyond superficial preferences for a given phone brand or a particular flavor of ice cream, beyond the ego-trip of believing we deserve to be a pop-star or the face on a celebrity poster. Resetting is not only about forgetting our habits or mundane preferences; it is about reconnecting to the ever-renewed origin of who we are, the qualitative feeling that constitute our unique experience of being alive: the process that we are as we exist, breathe, and sense our lifeworld.

Leibniz speaks of an identity of indiscernibles: if two things are utterly the same and occupy the same spacetime, they are one. All humans are discernible, and all the more so as their biography unfolds. We may discern someone else or ourselves in philosophical listening and calm observation or introspection. In his work *Oneself as Another* (1992), Ricoeur develops the concept of "narrative identity," the internalized and ever evolving story of the self that a person constructs to make sense and extract meaning out of their life. It allows us to understand ourselves through the mediation of otherness (situations, events, people, worlds…) while maintaining coherent selfhood. Otherness is not just projected objectively in a screen outside of our mind. It is a matter of individual perception, suggesting that recognizing alterity within the self is actually the path to authentic self-understanding rather than self-loss.

What resetting does is momentarily subtract ourselves from the realm that constitutes social and professional life. In the *cogito* (I think) or the *vivo* (I live), we are neither man nor woman, neither Chinese nor Canadian, neither rich nor poor, but simply capable of thought, feeling and being.

Amnesia, Poetry, and the Multiple Dimensions of Reset

The nineteenth century philosopher Nietzsche spoke of the importance of forgetting in order to achieve philosophical health. He did not mean forgetting everything, but unburdening the camel in us, creating space for the lion's freedom and eventually the child's innocent creativity. Heuristically speaking, a psychological analogy to a strong resetting might be amnesia: resetting would be, at least metaphorically, like an intentional, momentary, and selective form of amnesia (Kihlstrom, 2006).

A related idea is to use poetry as a tool for mental reset. Cognitive poetics investigates how literature and poetry can shift our usual modes of thinking by presenting unexpected and novel language and imagery (Gavins & Steen, 2003). Poetry can be like a shower for the mind and soul, an excellent way to reset cognitively by suggesting unexpected forms of language and images, like in a lucid dream. The poetic mind operates according to different rules than the analytical mind; it embraces paradox, celebrates ambiguity, and discovers meaning in the spaces between definitions (Hirshfield, 1997).

In a crealectic session setting, this can be implemented by reading or writing evocative poetry that disrupts habitual thought patterns. For example, Meena Kandasamy's poem "Amnesia, Selective" speaks of "unearthing events long past" and of nothingness, providing an ideal entry point into the resetting phase. Another—more radical—approach I used successfully in the past is to have participants write their full name on a piece of paper, and then delete one or two letters of it every time the facilitator pronounces one of the following sentences:

I do not know my real name…
I do not know my real age…
I do not know my real country…
I do not know my real language…
I do not know my real family…
I do not know my real planet…
I know that I am just born…

As an alternative to human poetry, nature can provide a profound setting for resetting. Walking in the forest near a dam overflow is an exercise I performed with some Vattenfall engineers (*vattenfall* means waterfall in

Swedish). In the "crealing" phase that succeeded the resetting phase (see Chap. 3), I told them to close their eyes and imagine that the concentrated sound of the water created by the spillway was the sound of the Creal, the creative cosmic onflow. Before this, in the spacetime of a clearing in the forest, during the resetting phase, I asked them to form a circle and I read out loud Robert Frost's "The Road Not Taken," triggering a multisensory resetting experience, combining the powers, sounds, and scents of the woods with poetic contemplation of life's bifurcations.

> Two roads diverged in a yellow wood,
> And sorry I could not travel both
> And be one traveler, long I stood
> And looked down one as far as I could
> To where it bent in the undergrowth;
>
> Then took the other, as just as fair,
> And having perhaps the better claim,
> Because it was grassy and wanted wear;
> Though as for that the passing there
> Had worn them really about the same,
>
> And both that morning equally lay
> In leaves no step had trodden black.
> Oh, I kept the first for another day!
> Yet knowing how way leads on to way,
> I doubted if I should ever come back.
>
> I shall be telling this with a sigh
> Somewhere ages and ages hence:
> Two roads diverged in a wood, and I–
> I took the one less traveled by,
> And that has made all the difference.

Resetting is about dimensions, the ones we see and the ones we feel, the ones we participate in and the ones we miss. I am my own dimension, and you are yours, each with sparks of love, admiration, ressentiment, and play. What are the dimensions in myself, emotional dimensions and intellectual dimensions that remain unfolded? What connections in the brain, what threads in the mind need to be untangled to later facilitate a rewiring?

These exercises and others help participants break free from their habitual frameworks and prepare their minds for the subsequent creative phases. In this book I intentionally focus more on the positive than the negative, but I sometimes think about those participating in a crealectic workshop that do not feel up to the creative and imaginative task; maybe they are ruminating a loss, a deception, a disease, a disappearance, their imagination hampered, closing themselves off from hope or fantasy. No dreamlike state, no multicolored pyramid of creation in the horizon. The resetting phase must be handled with care and sensitivity: it is not about forcing the blank slate, but about creating the conditions where genuine openness, gratitude, and even forgiveness become possible. Our occasional sadness and torment, when contemplated with detachment rather than ruminated upon, can be themselves a form of resetting, like listening to a musical lament on the radio.

Resetting is beginning to regenerate. It is vital for our health. Perhaps it's a de-setting rather than a resetting. A setting may be the place where we work, a familiar protocol of routines, ideas, ideologies even, ways of doing, familiar frames and judgements. Rather than moving from one setting to another setting, we may tend to explore the source of all settings, famously represented by the Construct in the original Matrix movie, a luminous infinite space that is like a virginal page in spacetime (and I sometimes show the Construct scene in the first Matrix movie as an exercise of resetting). The original setting of all settings is life, neither here nor there but everywhere at the same time, always here and now. In Sartre's existentialist framework, nothingness (*néant*) isn't merely the absence of being—it is a necessary condition for human consciousness and freedom (Sartre, 1956).

Beyond Autopilot: The Childlike Wonder

Admittedly, when we are faced with a project, problem, or challenge, we are prone to dive into our customary modes of thinking. This "autopilot cognition" (Kahneman, 2011) is our brain's quick way of dealing with routine tasks and decisions. This form of thinking is not inherently bad—it can be efficient for everyday tasks, but it becomes problematic when we need to innovate, feel alive and attentive or solve complex problems. The prevalence of cognitive biases demonstrates how our unreset minds can lead us astray. Behavioral economist Dan Ariely's research on predictably irrational decision-making reveals how our cognitive shortcuts, while often useful, can systematically distort our perception of possibilities (Ariely, 2008). Resetting involves temporarily suspending these automatic processes, creating space for a clear view on what Ariely calls "coherent arbitrariness," the recognition that many of our preferences and assumptions are more constructed and mechanical than we typically acknowledge.

These cognitive biases, a tendency to search for, interpret, and recall information that confirms our pre-existing beliefs, hinder our ability to think critically or creatively, and they distort our decision-making processes. By becoming aware of at least some of our presuppositions, we can strive to minimize their effects and open our minds to the possible. Csikszentmihalyi's research on "flow" underscores the importance of stepping back from our usual but somnolent patterns of thought to nurture a form of alert inspiration (Csikszentmihalyi, 1996). To be effective, this attitude needs to be practiced daily: it is not enough to engage in just a few annual "out-of-the-box" sessions. The crealectic reset must ideally become a habitual practice, a part of our professional, existential, and philosophical hygiene. Enterprise culture, for instance, should embrace compossible thinking and have resetting as a healthy protocol.

What would happen if we did not reset? We would perhaps treat creative exploration as impossible, or a banality, a routine itself, a problem-solving feature limited by our existing frameworks and short-term problems to fix. The risk of skipping the reset is the risk of stagnation, of becoming imprisoned in our own tired paradigms. Not everything needs to become conscious as we journey into the crealectic method. Our brain uses images, places, and faces as symbols of ideas. It activates messages indirectly, through associations and patterns that may not be immediately transparent to conscious awareness. This indirectness is not a flaw but a feature; it allows for connections and insights that purely logical, sequential thinking cannot access.

Resetting reconciles us with this indirect mode, this dreamlike state of mind where boundaries between concepts are fluid and new associations can form. Recent research in cognitive science has shown that mind-wandering, far from being an unproductive mental state, is crucial for creative thinking and problem-solving (Baird et al., 2012). We can be inspired by distraction. When we allow our minds to wander during a reset, we are not being idle, we are engaging brain regions associated with autobiographical planning, creative incubation, and self-referential thinking. These processes can lead to unexpected insights and novel connections that would be inaccessible through focused, analytical thinking alone. Serendipity occurs when we find something fruitful that we were not searching for. It is different from the streetlight effect, by which people search for something only where it is easiest to look. This universal joke is worth retelling: a policeman sees a man on his knees, apparently searching for something under a streetlight. The policeman asks: What have you lost? My keys, answers the man. They both look under the streetlight together. Are you sure you lost them here?, asks the policeman after a few vain minutes. No, replies the man, I lost them in the park. But why are you searching here then?, asks the policeman bewildered. The man looks at him with both anger and despair, and says: this is where the light is!

Philosopher Michael Bishop's investigation into well-being reveals that what he calls "psychological richness"—exposure to novelty, complexity, and perspective-changing experiences—is as important for human flourishing as happiness or meaning (Bishop, 2015). Resetting cultivates this psychological richness by deliberately disrupting our habitual patterns, creating what Bishop terms a space for "optimal novelty" that expands rather than overwhelms our cognitive capacities.

Resetting allows us to distance ourselves from deterministic processing and feel ready to engage in genuine creativity. It is what makes us more than biological computers executing programs. It is the seat of our dignity and agency. And it is this dignity and agency that the resetting phase aims to restore. Studies in contemplative neuroscience have shown that practices that cultivate this metacognitive awareness—this ability to observe our own mental processes—strengthen the prefrontal regions of the brain associated with executive function and creative problem-solving (Lazar et al., 2005; Siegel, 2007).

Children are often good at resetting. For instance, my son Theo, at 3 years old, told me this morning that "the sun is inside the moon." This statement, nonsensical to an adult mind locked in its certainties, reveals the fresh perspective of a mind not yet constrained by conventional knowledge. The moon does reflect the sun, absorbs and mimics its light. In the word "moon," the letters "o-o" could represent not only one but two suns (as a circle). There are many ways in which the sun could be "inside" the moon, some of them more rational, others more imaginative or poetic (you may make a list of ten of them as an exercise in resetting). Children constantly reset playfully because they have not yet built the elaborate cages of assumption, submission, and habit that adults live within. Their consciousness flows more freely, unimpeded by the accumulated debris of experience and framed socialization. They do not feel obliged to answer like a good soldier or follow an agenda scrupulously.

Developmental psychologists like Alison Gopnik have demonstrated that children possess heightened awareness in certain domains compared to adults, what she calls the "lantern" versus the "spotlight" of adult attention (Gopnik, 2009). While adults focus narrowly on what they deem relevant, children take in a broader spectrum of information without pre-filtering. This expansive awareness is precisely what we aim to achieve in the first steps of the crealectic method. The philosopher Gaston Bachelard wrote extensively about the phenomenology of reverie, a dreamlike state distinct from both directed thought and nocturnal dreaming. In *The Poetics of Reverie* (1960), he describes this state as one in which consciousness relaxes and wanders, allowing us to commune with the rhythms underlying existence. This reverie is not an escape from reality but a deeper immersion in it, accessing dimensions of experience that our practical, goal-oriented consciousness typically filters out.

The sun is inside the moon. In the resetting phase, we aim to recapture something of this childlike state—not to abandon reason or truth (the sun?), but to supplement it with wonder (the moon?). To see the world not as a collection of known quantities but as a field of marvels waiting to be unveiled. This state of wonder and admiration is the fertile soil in which creative ideas germinate. Admiration was for Descartes the mother of all emotions. The ability to transform the familiar into the strange, to see everyday objects, processes, and problems as if for the first time, lies at the heart of philosophy and creativity across disciplines (Root-Bernstein & Root-Bernstein, 1999).

The Romantic poet Shelley described poetry precisely as making the familiar strange and the strange familiar. Einstein famously used thought experiments to reset his perspective on physical problems, imagining himself riding alongside a beam of light. Darwin reset scientific thinking about species by refusing to see them as fixed categories. In industrial settings, we often see the best innovations begin with a reset, a willingness to set aside established ways of perceiving and categorizing the world.

The Antechamber of Rewilding and Deworlding

Resetting often inhabits a mental and emotional dimension that transcends purely rational attitude. It invites us into a space where logic loosens its grip, where the tyranny of usefulness is suspended, where play and imagination can flourish without immediate purpose. This dimension is not, as many tend to think, facultative; it is functional at the deepest level, enabling modes of cognition that analytical thinking alone cannot access.

The poet John Keats spoke of "negative capability," the capacity to remain in uncertainties, mysteries, and doubts without irritably or fearfully reaching after fact and reason. This capacity is a deliberate suspension of our rush to conclusion, our reflex to categorize and judge, our habit of filtering experience through preestablished frameworks. When we reset, we dwell temporarily in this space of not-knowing, and from this fertile void, new possibilities may later emerge.

Gary Snyder, poet and environmental activist, describes his attention to a "practice of the wild," a form of resetting that involves reconnecting with the untamed aspects of consciousness as well as nature. "Wild mind," he suggests, is not alienated, but "elegantly self-disciplined, self-regulating. That's what wilderness is. Nobody has a management plan for it" (Snyder, 1990). This wildness—this unmanaged craft of mind—is what we seek to welcome in the first three stages of crealectic thinking (Resetting, Crealing, Profusing), to allow for the higher order of the last two stages (Compossibilizing, Realizing).

In his phenomenological investigations, Bachelard argues that reverie "gives us the world of worlds. Poetic reverie is a cosmic reverie. It is an opening to a beautiful world, to beautiful worlds. It gives the I a non-I which belongs to the I: my non-I" (Bachelard, 1960). This paradoxical formulation captures something essential about the resetting process; it introduces a semi-otherness into our consciousness, a natural freshness that is only superficially alien, an untamed capacity to see from outside our habitual perspectives while being more authentically ourselves. This is not an abandonment of self but an expansion of it, a reconnection with dimensions of our being that have been obscured by habit and narrowly focused attention. This is the space that our institutional or corporate roles neglect to explore because they are not administratively coded.

Anthropologist Tim Ingold speaks of the importance of what he calls "attention to the possible," learning to attend to the world in new ways (Ingold, 2022). This attentive restoration means forsaking the certainty of knowledge that categorizes and explains, instead embracing the vulnerability and astonishment that comes with truly attending to reality. Rather than seeing attention as merely confirming what we already know, we must recognize that our habitual ways of knowing often blind us to the fullness of presence. As Ingold suggests, "Blinded by knowledge, we can often fail to see what is before our eyes."

The anthropological concept of defamiliarization—making the familiar strange to see it freshly—is central to resetting. Victor Shklovsky, the

Russian formalist, called this process "ostranenie," a neologism designating a voluntary estrangement of discourse: "The purpose of art is to impart the sensation of things as they are perceived and not as they are known" (Shklovsky, 1965). This principle extends beyond art into all creative thinking. In order to reset, we must deliberately defamiliarize our relationship to our subject matter, our problems, our goals, and even our own identities. This can be done without sacrificing our philosophical health or our composure.

Cultural anthropologist Mary Douglas observed that impurity is sometimes just "matter out of place," a displacement that violates our categorical systems rather than something necessarily unclean (Douglas, 1966). This insight is an example of how our conceptual frameworks determine what we perceive as orderly or disorderly, relevant or irrelevant, signal or noise. Resetting involves suspending or bracketing these frameworks, not necessarily to invalidate them but to give peace and novelty a chance. It's not only about blankness and emptiness, it is also allowing us to perceive the "dirt" that might contain valuable information or perspectives we have been trained to filter out.

In the context of innovation and creative problem-solving, this anthropological reset may allow us to question assumptions that are so deeply embedded in our cultural and professional environments that they have become invisible to us. It helps us identify what Edward T. Hall called the "silent language" of culture—the implicit rules, values, and patterns that shape our thinking without our awareness (Hall, 1959). By making the silent principles of corporate or institutional culture intentional, and therefore audible, we create the possibility of working with them more consciously.

Edmund Husserl's phenomenological method offers a philosophical precedent for resetting through what he called "epoché" or "bracketing." In this practice, Husserl proposed suspending or "putting in brackets" our natural attitude—the taken-for-granted assumptions we normally hold about the world—to access phenomena as they present themselves to consciousness (Husserl, 1913/1982). This deliberate suspension of judgment doesn't deny the existence of the external world but temporarily sets aside our presuppositions about it, allowing us to encounter experience with fresher attention. Husserl's bracketing, which is a philosophical version of what later psychologists called mindfullness, represents a systematic approach to achieving what the crealectic method calls resetting: creating an open space of awareness where habitual frameworks are temporarily suspended. Like Descartes' cogito, Husserl's epoché offers a path to a more primordial experience, one that precedes our conceptual categorizations and naturalized beliefs. Through this tentative practice, we don't abandon the world but return to it with renewed perception, seeing beyond the sedimented layers of interpretation or judgement that normally obscure direct experience. This phenomenological reset doesn't just clear mental space; it transforms, when repeated, our relationship to experience itself, potentially revealing what Husserl called the "things themselves" beneath our habitual constructions. These things are not objects, but intercreative relationships in dynamic configuration. Here again, we should not be too perfectionist about the practice of resetting: this is not yet the realm of exactitude that will be achieved in compossibilizing and realizing. And we are not trying to reach a nirvana of nothingness to remain still in it. Resetting is a land we travel through, not a destination.

Cultural historian Wolfram Eilenberger's study of the philosophical revolution between 1919 and 1929 reveals how thinkers like Heidegger, Wittgenstein, Benjamin, and Cassirer each performed radical "resets" of philosophical thinking in response to civilizational crisis (Eilenberger, 2020). Their work demonstrates how periods of breakdown can become opportunities for fundamental reconceptualization, suggesting that our current challenges similarly call for philosophical reset rather than merely technical solutions. Beyond narratives of apocalyptic doom or business-as-usual continuation, the Creal flux persists as continuity rather than rupture.

The Neuroplasticity of Reset: Biological Foundations

The practice of resetting is not merely a philosophical or psychological exercise; it has concrete biological effects and origins. Neuroscientific research increasingly confirms that our brains retain remarkable plasticity throughout life. The neural networks that encode our habitual ways of perceiving and responding to the world can be reshaped through deliberate practice (Doidge, 2007).

When we engage in resetting practices, we are somewhat rewiring our brains. Studies of experienced meditators show significant differences in brain structure and function compared to non-meditators, particularly in regions associated with attention, sensory processing, and self-awareness (Davidson & Lutz, 2008). Similar changes have been documented in people who engage in other forms of mental training that involve stepping back from habitual patterns of thought and perception.

Recent research in predictive processing theory suggests that our brains are constantly generating predictions about what we may experience (Clark, 2013). This creates a self-reinforcing loop where we increasingly perceive what we already expect to perceive. Resetting disrupts this loop, allowing novel information and patterns to reach conscious awareness.

The practice of resetting also affects our autonomic nervous system, shifting us from the sympathetic "fight-or-flight" mode that narrows attention and prioritizes quick, habitual responses, to the parasympathetic "rest-and-digest" mode more conducive to expansive awareness and creative connection-making (Porges, 2011). This physiological shift creates the conditions for insights and novel associations to emerge. More precisely, creative states often involve what researchers call relaxed attention, a state that's neither high sympathetic arousal nor deep parasympathetic rest, but rather a balanced eudynamia that allows for flexible attention switching between focused and diffuse modes.

The neuroscience of creativity suggests creativity involves enhanced cooperation and integration between networks that normally work in opposition. (Beaty et al., 2018). But a temporary disinhibition is indeed needed to allow for the control system to renew its assumptions and normally suppressed or "noisy" neural networks to communicate. This is facilitated by resetting practices, making space for unusual connections to form between ideas and experiences normally kept separate in our categorical thinking.

Practical Applications: Resetting in Organizational Contexts

At Vattenfall and other organizations where I have applied the crealectic method in workshops, resetting has proven to be well accepted by participants. Most participants who volunteered their presence did not seem to resist the invitation to step outside their professional roles and expertise, to temporarily set aside the knowledge and skills that define their value within the organization. One senior engineer remarked: "I realized I had been looking at the energy distribution problem with the same mental model for fifteen years. During the first hour of the crealectic workshop, I saw three alternative approaches I had never considered."

The efficacy of resetting in organizational contexts depends on creating what psychologist Amy Edmondson calls "psychological safety," an environment where people feel they can take interpersonal risks without fear of punishment or humiliation (Edmondson, 1999). Without this safety, participants may go through the motions of resetting without relinquishing any of their habitual defenses and thought patterns. Facilitators must therefore establish a diplomatic attitude that validates the vulnerability inherent in authentic resetting.

Organizational theorist Karl Weick has written extensively about the importance of "dropping our tools"—literally and metaphorically releasing the implements and frameworks that define our professional identities—in order to reflect on changing circumstances (Weick, 1996). He cites the tragic example of firefighters who died because they refused to drop their heavy equipment when fleeing a rapidly advancing fire. In less dramatic but equally consequential ways, professionals in all fields often cling to their conceptual tools even when these frameworks are not serving them anymore. Resetting provides a structured practice for temporarily dropping these tools to see situations freshly.

Ten Exercises for Resetting

The following exercises for resetting, some of which I already referred to, can be adapted to various contexts and personal preferences. These are not mere techniques but invitations to a different quality of consciousness, portals through which we might make ourselves available to access the creative wellspring of the Creal, which is the topic of the next stage in the method:

1. The Cogito Meditation: Spend five minutes doubting everything except the fact that you are aware of thinking. Follow Descartes in his radical doubt but preserve the certainty of your own consciousness as consciousness. Feel yourself as pure awareness and thinking-feeling subject, prior to all accumulated identities and beliefs. This practice dissolves some rigid structures of ordinary thought, allowing the fluid potentiality of the *I am* to emerge in consciousness.

2. Poetry Immersion: Read, write, or listen to a poem or text that uses unexpected imagery and language. Allow the words to wash over you without analyzing them intellectually. Let Wallace Stevens take you to "The House Was Quiet and the World Was Calm," or Rilke remind you that "You must change your life." Poetry bypasses our analytical filters, speaking directly to the pre-conceptual awareness where reset occurs. As long as the poem irritates you, you may not be yet resetting peacefully (or perhaps it's a bad poem). In the context of group workshops, we may ask each participant to write a verse unbeknown to the others, thus creating a multiversal collective poem that we can then read to the intercreative audience.

3. "Empty the Room": Close your eyes, then gradually erase people and objects in the room in which you are seating or standing. With each chair removed, each person deleted, feel yourself closer to the luminal emptiness that is the canvas of reality. Automatically, you will start to breathe deeply and feel relaxed, in a state of gracious reverie. This relaxing exercise has proven to work very well and easily so, both in group gatherings and individual context.

4. Nature Absorption: Move to a natural setting—a park, a river, a forest—and concentrate fully on one element (the sound of leaves, the pattern of water, the texture of bark) for five minutes. Nature follows its own rhythms, somewhat indifferent to our human categories and concerns. By briefly attuning to these rhythms or what we may perceive of them, we temporarily step outside the constructed world of human meaning into a direct experience of what phenomenologists call "the things themselves" and psychologists call "mindfulness."

5. Sensory Focusing: Close your eyes and focus exclusively, but not eagerly, on one sense (hearing, touch, smell) for several minutes, noting details you normally overlook. Our perceptual systems are designed to filter out information deemed irrelevant to our practical concerns. By deliberately attending to normally backgrounded sensory data, we disrupt these filters and access a richer, more textured experience of reality, or at least distinguish the relativity of what seemed natural or normal.

6. Beginner's Mind: Choose a familiar object and try to examine it as if you've never seen it before. What would a being from another planet make of this object? The Zen concept of *shoshin* or "beginner's mind" suggests that renewed insight may come when we abandon our expert knowledge and attempt to approach experience with fresh, unhabituated attention. This practice helps us see beyond our conceptual overlays to the vibrant specificity of things considered as possibilities.

7. Physical Reorientation: Change your physical position dramatically—lie on the floor instead of sitting, stand on one leg, or invert your usual posture—while contemplating a problem or question or with no particular thought in mind. Our thinking is embodied; changing our physical relationship to space alters or pauses our cognitive patterns. This practice leverages what philosopher Alva Noë (2004) calls "enactive perception," the insight that perception is not passive reception but active exploration through bodily engagement with the environment.

8. Reverse Assumptions: Take three fundamental assumptions about your field, project, or problem and deliberately reverse them into an opposite stance. What possibilities emerge? This practice builds on what psychologist Ellen Langer calls "sideways learning," approaching situations from unexpected angles rather than following the well-worn cognitive paths of habit (Langer, 1997). It temporarily frees us from what philosopher Alfred North Whitehead (1978) called "the fallacy of misplaced concreteness" (of which more in the next chapter), mistaking our useful categories for reality itself.

9. Silence Immersion: Practice five minutes of complete silence, focusing only on the spaces between your thoughts rather than the thoughts themselves. The Daoist tradition speaks of wu-wei or "non-doing," an effortless action that arises from alignment with the natural flow of things rather than from egoic striving. This practice cultivates the receptive attentive ness that allows us to hear what wants to emerge beyond our preconceptions, or in some cases, it gets us closer to a fecund daydreaming state.

10. Artistic Disruption: On a given theme (each crealectic workshops should have a theme, not negatively formulated as a problem to solve but rather as an ideal to explore), create a quick, spontaneous drawing, collage, dance, sentence or melody that describes how you have experienced or thought of it until now. Don't aim for beauty or coherence: aim for free association, spontaneity and play. This practice embodies what anthropologist Gregory Bateson called "the difference that makes a difference," creating a meaningful disruption in our habitual patterns that can serve as the seed of future *in-formation (1972)*.

These exercises, merely indicative (the list is infinite), are not mere distractions or warm-ups, even if they can be experienced joyfully. They are deliberate interventions designed to disrupt habitual thinking and create the mental space necessary for genuine innernovation. They represent different doorways into the same essential shift in consciousness, from identifying with our accumulated knowledge and established patterns to abiding in the open awareness that is our active birthright.

Through the practice of resetting, we prepare ourselves for the next phase of the crealectic method: crealing, where we reconnect with the Creal, the creative flow of potential that underlies all reality.

The transition from resetting to crealing is not abrupt but fluid. As we clear away the mental debris or scaffolding of habit, we naturally begin to sense the vibrant, generative current of live pulsing beneath the surface of ordinary experience. But without the blanker slate provided by resetting, this reconnection with life itself would be too abrupt, hindered by our preconceptions and perhaps limited by our expectations.

"The mind," wrote Plutarch, "is not a bottle to be filled, but a fire to be kindled." Resetting is not about emptying ourselves of all content in order to be filled with totally new content, but about rekindling the fire of original awareness that may burn away our calcified certainties and illuminate new possibilities. It's about remembering what we did not know before we unlearned to forget, while reconnecting with our impulse to think independently and our ardent desire for the truth behind the accepted norms.

Reality is not a static collection of defined objects but a dynamic play of intercreative becoming. Now available to ultimate possibility after resetting, we are ready to participate more consciously in the great unfolding of *crealia*, the creative inflorescences that are both our deepest nature and the nature of the cosmos itself.

References

Ariely, D. (2008). *Predictably irrational: The hidden forces that shape our decisions.* HarperCollins.

Bachelard, G. (1960). *The poetics of reverie: Childhood, language, and the cosmos.* Beacon Press.

Baird, B., Smallwood, J., Mrazek, M. D., Kam, J. W., Franklin, M. S., & Schooler, J. W. (2012). Inspired by distraction: Mind wandering facilitates creative incubation. *Psychological Science, 23*(10), 1117–1122.

Bateson, G. (1972). *Steps to an ecology of mind: Collected essays in anthropology, psychiatry, evolution, and epistemology.* University of Chicago Press.

Baumeister, R. F., Bratslavsky, E., Muraven, M., & Tice, D. M. (1998). Ego depletion: Is the active self a limited resource? *Journal of Personality and Social Psychology, 74*(5), 1252–1265.

Beaty, R. E., Kenett, Y. N., Christensen, A. P., Rosenberg, M. D., Benedek, M., Chen, Q., & Silvia, P. J. (2018). Robust prediction of individual creative ability from brain functional connectivity. *Proceedings of the National Academy of Sciences, 115*(5), 1087–1092.

Berman, M. G., Jonides, J., & Kaplan, S. (2008). The cognitive benefits of interacting with nature. *Psychological Science, 19*(12), 1207–1212.

Bishop, M. A. (2015). *The good life: Unifying the philosophy and psychology of well-being.* Oxford University Press.

Clark, A. (2013). Whatever next? Predictive brains, situated agents, and the future of cognitive science. *Behavioral and Brain Sciences, 36*(3), 181–204.

Csikszentmihalyi, M. (1996). *Creativity: Flow and the psychology of discovery and invention*. Harper Perennial.
Davidson, R. J., & Begley, S. (2012). *The emotional life of your brain: How its unique patterns affect the way you think, feel, and live – And how you can change them*. Hudson Street Press.
Davidson, R. J., & Lutz, A. (2008). Buddha's brain: Neuroplasticity and meditation. *IEEE Signal Processing Magazine, 25*(1), 176–174.
Descartes, R. (1999). *Discourse on method and meditations on first philosophy*. Hackett.
Doidge, N. (2007). *The brain that changes itself: Stories of personal triumph from the frontiers of brain science*. Viking.
Douglas, M. (1966). *Purity and danger: An analysis of concepts of pollution and taboo*. Routledge.
Edmondson, A. (1999). Psychological safety and learning behavior in work teams. *Administrative Science Quarterly, 44*(2), 350–383.
Eilenberger, W. (2020). *Time of the magicians: Wittgenstein, Benjamin, Cassirer, Heidegger, and the decade that reinvented philosophy* (S. Whiteside, Trans.). Penguin Press.
Gavins, J., & Steen, G. (Eds.). (2003). *Cognitive poetics in practice*. Routledge.
Gopnik, A. (2009). *The philosophical baby: What children's minds tell us about truth, love, and the meaning of life*. Farrar, Straus and Giroux.
Hall, E. T. (1959). *The silent language*. Doubleday.
Hirshfield, J. (1997). *Nine gates: Entering the mind of poetry*. HarperCollins.
Hunter, E. M., & Wu, C. (2016). Give me a better break: Choosing workday break activities to maximize resource recovery. *Journal of Applied Psychology, 101*(2), 302–311.
Husserl, E. (1982). *Ideas pertaining to a pure phenomenology and to a phenomenological philosophy: First book* (F. Kersten, Trans.). Kluwer Academic Publishers. (Original work published 1913).
Ingold, T. (2022). On not knowing and paying attention: How to walk in a possible world. *Irish Journal of Sociology, 31*(1), 20–36.
Kahneman, D. (2011). *Thinking, fast and slow*. Farrar, Straus and Giroux.
Kihlstrom, J. F. (2006). Trauma and memory revisited. In B. Uttl, N. Ohta, & A. L. Siegenthaler (Eds.), *Memory and emotion: Interdisciplinary perspectives* (pp. 259–291). Blackwell.
Knapp, J., & Zeratsky, J. (2018). *Make time: How to focus on what matters every day*. Currency.
Langer, E. J. (1997). *The power of mindful learning*. Addison-Wesley.
Lazar, S. W., Kerr, C. E., Wasserman, R. H., Gray, J. R., Greve, D. N., Treadway, M. T., & Fischl, B. (2005). Meditation experience is associated with increased cortical thickness. *Neuroreport, 16*(17), 1893–1897.

Levitin, D. J. (2014). *The organized mind: Thinking straight in the age of information overload*. Dutton.

Luders, E., Toga, A. W., Lepore, N., & Gaser, C. (2009). The underlying anatomical correlates of long-term meditation: Larger hippocampal and frontal volumes of gray matter. *Neuroimage, 45*(3), 672–678.

Newport, C. (2016). *Deep work: Rules for focused success in a distracted world*. Grand Central Publishing.

Noë, A. (2004). *Action in perception*. MIT Press.

Porges, S. W. (2011). *The polyvagal theory: Neurophysiological foundations of emotions, attachment, communication, and self-regulation*. W. W. Norton.

Ricoeur, P. (1992). *Oneself as another*. Chicago University Press.

Root-Bernstein, R., & Root-Bernstein, M. (1999). *Sparks of genius: The thirteen thinking tools of the world's most creative people*. Houghton Mifflin.

Sartre, J. P. (1956). *Being and nothingness*. Philosophical Library.

Shklovsky, V. (1965). Art as technique. In L. T. Lemon & M. J. Reis (Eds.), *Russian formalist criticism: Four essays* (pp. 3–24). University of Nebraska Press.

Siegel, D. J. (2007). *The mindful brain: Reflection and attunement in the cultivation of well-being*. W. W. Norton.

Snyder, G. (1990). *The practice of the wild*. North Point Press.

Stuckey, H. L., & Nobel, J. (2010). The connection between art, healing, and public health: A review of current literature. *American Journal of Public Health, 100*(2), 254–263.

Weick, K. E. (1996). Drop your tools: An allegory for organizational studies. *Administrative Science Quarterly, 41*(2), 301–313.

Whitehead, A. N. (1978). *Process and reality*. Free Press.

CHAPTER 3

Crealing

Abstract Crealing represents reconnection with the creative flux underlying all reality—the Creal (Creative Real). This chapter examines how embodied engagement with cosmic creativity differs from intellectual analysis, exploring connections to process philosophy, autopoietic systems, and quantum physics. It addresses obstacles to creative connection and provides practical approaches for individuals and groups to access and maintain awareness of their participation in more-than-human creative processes, emphasizing felt experience over blinded productivism.

Keywords Crealing • Creative Real • Embodied cognition • Quantum consciousness • Autopoiesis • élan vital • Participatory consciousness • Flow states • Creative emergence • Proprioception

L. de Miranda, *Crealectics as a Creative Method*,
Palgrave Advances in Possibility Studies,
https://doi.org/10.1007/978-3-032-00427-7_3

Fig. 3.1 The crealing phase symbol

We walk into the shore where the ocean meets the land, no longer as observers but as participants, ready to dive anew into the sea of ultimate possibility. The waves soon echo the waving of our arms, inviting us to wade deeper. In this chapter, we will ascend to the heights of cosmology before descending back to the fields of practice.

In the previous chapter, we explored resetting, the conscious clearing of mental space, a walk through the sand beach in which all grains are like atoms of reality. With minds freshly available, we turn to the second phase of the crealectic method: *Crealing* (Fig. 3.1). The term crealing—a blend of "creation" and "feeling" based on the concept of Creal or Creative Real—invites us to reconnect with the creative rhythm and whole onflow that pulses not only within our bodies but throughout all reality. This constant ebb and flux, this perpetual intercreation of life and explosion of cosmic stuff, echoes what twentieth century philosopher Henri Bergson termed the "creative élan vital" (Bergson, 2010), the life force that drives evolution and operates as the ultimate dynamo of possibilities.

The Creal operates not merely as flowing élan vital but sometimes as what we might call "élan explosif"—a creative force that doesn't simply flow but also detonates into new possibilities. Each moment may be a droplet in a stream or a micro-explosion in an endless chain action and reaction of becoming.

The Ocean of Creation: Reconnecting with the Creative Source

At the heart of life as a felt becoming, one discovers the ocean of creation and *eudynamia* ("good potential"). In a way, the pre-Socratic philosopher Thales intuited a profound truth when he proposed that water was the *archē*, the fundamental principle of reality. Not H_2O, but rather the fluid flux of potential that predates, overflows, and initiates the apparent stability we call world. We are swimmers, divers, fish in this all-encompassing flux of potential, of which we as persons perceive only a fragment—much like dreamers perceive only aspects of the dreaming multiverse.

What is the Creal? It is a force, a flow, a ferocious kindness, a glorious impetus, never ceasing its metamorphosis, weaving the One and the Multiple into bursts of possibility and proposals of compossibility. It is the creative engine of the universe, one in which creation is also an emotion. When we creal, we reconnect with the emotion of emergence by which wholes and parts are in constant dialogue. In crealing we pause our utilitarian calculations to welcome the sublime gratitude to the Ultimate, whatever me may or may not call it (Dao, God, Nature or Pachamama). To truly grasp the essence of crealing, we need to appreciate the nuanced interplay between our physical being and the larger universe. This is not only observation but participation in the manifold of possible worlds. "Each portion of matter may be conceived as a garden full of plants, and as a pond full of fish. But each branch of the plant, each member of the animal, each drop of its humors, is also such a garden or such a pond" (Leibniz, 1989). Each drop of water is a lifeworld in itself, a micro-universe, wrote Leibniz in his *Monadology*. Multiverses constituting worlds of compossibility, and their vastitude, are what we see when we raise our heads from our pettyfogging concerns.

As physicist David Bohm (1980) explains, the world which we perceive cannot be separated from the perceptual process; the observer is both participant and onlooker. Heisenberg (1999) noted: "What we observe is not nature itself, but nature exposed to our method of questioning," reminding us that we are always intercreating with the universe rather than merely observing it from outside. Process philosophers have rediscovered the importance of seeing the world not as inert substances manipulated mechanically at a distance, but as perspectival intertwinements (Barad, 2007; Smolin, 2013; Whitehead, 1978). The sense that creativity is a ubiquitous process ongoing all around us, connecting all subjects, beings, and entities in dynamic systems, was often present in ancient cosmologies around the world; today, it comes anew from various angles, among which posthumanism (Braidotti, 2013; Haraway, 2016), new materialism (Coole & Frost, 2010), and the practical cosmology of crealectics (de Miranda, 2017).

Science offers inspiring insights into this understanding. Quantum mechanics has famously revealed how our observations influence the outcome of events at the subatomic level (Penrose, 1989). This entanglement of observer and observed underscores that we are intrinsically linked with the ongoing processes around us. But in crealing, we do not only observe or manipulate: we augment our creative energy, our creative joy, our sense of the possible. Healthy power is this sense of the possible, the power to participate consciously in bringing forth new compossible realities. The Creal should not become a dogma but remain a fecund hypothesis. Neuroscientific research shows that engaging in creative inspiration does stimulate brain regions associated with problem-solving and emotional regulation (Dietrich & Kanso, 2010). By tapping into the felt act of continuous creation, we are also reaping tangible cognitive and emotional benefits.

The Wave Function Underlining Reality

The Creal can be metaphorically compared to the universal wavefunction of quantum physics, the field of all possibilities before their collapse into specific and narrowed actualities. We live among the shadows of collapsed waves, and these possibilities appear and die. But the ocean of potential remains ever-present, ready to manifest in new forms through our creative engagement. In crealing, we aim to de-collapse the wave function, to recreate in the mind a quantum fluctuation that reveals the ultimate potential hidden beneath the apparent solidity of the material world.

As twentieth century quantum physicist Erwin Schrödinger (1992) suggested in his groundbreaking work *What Is Life?*, originally published during the Second World War, sometimes a new way of understanding reality comes from comparing two fields that are not usually in dialogue, like science and intuition. Crealing invites us to see the quantum nature of reality in our everyday intuitive experience of life, to perceive the vibrant probability fields ("*crealia*") that constitute even the most ordinary objects around us. Creation is usually perceived as an act of production; what if it is originally an emotion?

What crealing does is take us to the core of our capacity to possibilize, which twentieth century psychologist Ludwig Binswanger, Freud's intimate friend, recognized as essential to mental health. Binswanger spoke of "re-possibilisation" for those who have lost their capacity to feel the possible and have a hindered sense of potential. Crealing, when repeated as a practice of attunement with life itself, reactivates our sense of the possible, augments it, makes it healthy again. This is an antidote to what philosopher Mark Fisher (2009) called "capitalist realism," the pervasive sense that no alternative to the current social and economic system is possible or even imaginable. How do we go back to possibilizing? Most systems of order may request agents to be somewhat depossibilized, to function predictably within established parameters, to be not only politically but statistically correct. What would happen in our world if everyone were a compossibilizer? We would all engage consciously in the work of crealing, the deciphering of reality's conventional structures to reveal the creative magma beneath, not one of chaos only but also of eudynamic harmony in the making. This involves momentarily surrendering analytic intelligence, even dialectic intelligence, and embracing crealectic intelligence.

Some say that consciousness collapses the wave function, but one could argue that in crealing consciousness reawakens the wave function, transforming the Real back into Creal through the alchemy of felt and grateful attention. The Creal can be understood as this very gratitude, a loving rediscovery of hidden superabundance via the courage to abandon our realistic certainties and sense of limitation. In crealing we daydream, not only connecting with the sense of childlike wonder but also diving into the participatory aspect of that wonder.

Possibility is often subordinated to actuality in conventional thinking. We privilege what exists over what might exist, the concrete over the potential. But this hierarchy is arbitrary and limiting. The possible is not less real than the actual; it is simply awaiting manifestation. As process philosopher Gilles Deleuze argued, the virtual (his term for the realm of pure potential) is fully real in itself, not merely a shadow of the actual (Deleuze, 1994).

The dominance of the actual over the possible in our everyday realistic thinking manifests sometimes in economic discourses that paint Earth as a rock of scarcity rather than an ocean of abundance (Glaveanu & de Miranda, 2021). Capitalism frames existence as lacking and encourages us to build over emptiness, bridge over nothingness, and productively compete for supposedly scarce resources. The capitalist discourse disciplines the possible by never ending its process of production, covering emptiness as if it were a land to colonize, pretending to fear scarcity in order to mobilize our efforts. But there is nothing to fear where abundance is given in the form of ultimate potential rather than things to purchase, consume or exhaust.

Embodied Crealing

Most people have access to crealing via aesthetic experiences—musical joy, artistic activities, natural beauty. These experiences provide what philosopher Martin Heidegger called "clearings" (*Lichtungen*), openings in the dense forest of everyday concerns where Being can reveal itself more directly as the dynamic verb of life, indeed also a becoming. The felt sense of continuous creation often manifests as physical sensations: tingles along the spine when hearing powerful music, the catch in the throat before tears of beauty, the expansive relief in the chest during moments of awe in nature. These embodied responses are not mere epiphenomena but direct apprehensions of our participation in the sublime.

Contemplative teacher Richard Rudd distinguishes between "concentration" (focused attention on objects) and "contemplation" (open awareness of the creative source itself) (Rudd, 2018). Crealing corresponds to this contemplative mode, what Rudd describes as "merging with the field of transformation," the space between thoughts where new possibilities spontaneously emerge as an art of living wakefully.

Crealing invites us to feel rather than merely intellectualize the creative source that supports life and possibility. We can experience an embodied apprehension of life as an intercreative reality, a network of ecologies, where humans are an integral part of the natural world, not separate from it (Spinoza, 1996). Feeling this creative flux reveals our sense of the possible and of the potential harmony between possibles, broadening our understanding and capabilities. Henri Bergson's investigations into the nature of creative evolution distinguish between what he calls "intellect," which spatializes and fragments experience, and "intuition," which grasps the flowing continuity of creative becoming (Bergson, 2010). Crealing cultivates this intuitive mode, allowing us to participate in what Bergson, inspired by Darwin, describes as the creative impulse that drives all evolutionary development.

Philosopher Maurice Merleau-Ponty (2012) understood this embodied form of knowing, arguing that our bodies are not objects in space but the very medium through which we have a world at all. Crealing embraces this carnal epistemology, this knowing-through-embodiment that precedes and grounds all abstract cognition. As neuroscientist Antonio Damasio (1999) has demonstrated, emotions are not interruptions of reasoning but essential components of it, providing the somatic markers that guide our decisions and judgments.

Crealing also involves acknowledging and overturning our disconnection from the natural world, developing a sense of oneness with Earth as a complex, self-regulating system (Lovelock, 1979). This process invites us to self-transcend, or rather to "inscend" within our immanent nature (Berry, 2002). It is the realization that we are never fully separate from anything else, that the boundaries we perceive are often useful conventions rather than absolute divides. Recent research in embodied cognition confirms what phenomenologists have long argued: that our thinking is not disembodied computation but fundamentally grounded in our physical experience of the world (Lakoff & Johnson, 1999; Varela et al., 1991). We are embedded in societies and extended by the world. The evolutionary body is not a container for consciousness but the very medium through which consciousness emerges and operates. When we creal, we engage this embodied intelligence, allowing somatic wisdom to inform our creative process. We reconnect with the long duration of cosmic time, and the spacetime of evolutionary transmutation by which we are connected with all forms of life in the past, present, and future. This manifests in what philosopher Eugene Gendlin called the "felt sense" that precedes and guides conceptual thinking (Gendlin, 1978).

The Creal energy nurtures us like any material food, if not more. But it's like oxygen or love: we tend to only think about it when it withdraws in decompression or depression. There is an eternal recycling of the Real in the Creal, but we may also hypothesize an emergence from nothingness, a *creation ex nihilo*. There may be infinite being in nothingness itself, as Asian mystics have proposed. We feel the infinite roads not taken. Bifurcations of joy or apprehension.

As philosopher Ernst Bloch (1986) argued in his monumental work *The Principle of Hope*, human existence is characterized by a fundamental orientation toward the "not-yet," toward possibilities that have not been realized but draw us forward in anticipation. This "anticipatory consciousness" is not merely wishful thinking but a vital element of human creativity and social transformation. Crealing activates this anticipatory consciousness, helping us to perceive and engage with the "not-yet" that pulls us toward new horizons of possibility, but also to the origin of this not-yet.

The Creal contains the Real. Reality can be a virtual source for another reality, a possible for another possible. But the creative emergence might happen in a different world. There is a plurality of lifeworlds on planet Earth. Each living being is in a certain way a world in itself. Earth is a multiverse of perspectives.

The Autopoietic Universe: Crealing as Playful Self-Organization

Biologists Humberto Maturana and Francisco Varela introduced the concept of autopoiesis to describe the self-creating, self-organizing capacity of living systems (Maturana & Varela, 1980). The universe itself displays this autopoietic quality, continuously regenerating and reorganizing itself through creative processes. When we creal, we tap into this fundamental cosmic poiesis—explosive and unifying self-creation that shatters existing forms to generate radically new ones.

There is a resonance between this biological concept and what twentieth century mathematician and philosopher Alfred North Whitehead (1978, p. 21) called "creativity," which he described as "the universal of universals characterizing ultimate matter of fact." For Whitehead, creativity is not something humans add to a static reality but the very essence and matter of reality itself, "the principle of novelty" through which the many may become one and are increased by one.

The poet Rainer Maria Rilke, in his *Letters to a Young Poet* (2004), beautifully describes how creativity works as a natural organizing process. He suggests that everything that affects us will naturally find its proper place in our understanding. This happens because, beyond intellectual knowledge, we're entitled by nature to a deeper experience that encompasses virtually everything. When we engage in creative acts, we're participating in this natural process, allowing our impressions and ideas to arrange themselves meaningfully in both our understanding and expression, with time or within a lively duration. Writing, composing, or thinking authentically is part of the autopoiesis of our biography.

Creation creates as nature "natures"—through a more or less spontaneous self-organization that requires sometimes minimal intervention from an external designer or imposed plan. In crealing, we're not imposing structure on chaos but allowing the natural eudynamic organizing tendencies of the Creal to express themselves through us. As Kauffman writes (1993), "Life exists at the edge of chaos, poised between too much order and too much disorder." Connecting thought with the process we call life enables us to see beyond the given, to see alternatives as they unfold, to discover new connections and combinations that emerge from sense-making while constituting meaning as order. Knowing that we do not need to generate 100% of the world order via effort allows us to transcend the limitations of our current understanding and engagement with reality. Crealing is also deep listening to what the earth is whispering in terms of possible configurations.

Many spiritual traditions emphasize the playful, joyful quality of cosmic creativity. In Hindu philosophy, the concept of Lila describes the divine play of creation, the delight that underlies all becoming. The Upanishads speak of creation as *Ananda*, bliss. This life-affirming drive is not incidental but essential to understanding the nature of reality. Creation is not grim necessity but joyful expression, not stern duty but exuberant play. But it is also an awareness of the trauma that runs through all earthling experience, the wound of imagining ourselves separated from the Oneness of Creal.

Heraclitus famously said that "eternity is a child playing, moving counters in a game." This image of the cosmic child at play captures something essential about crealing—its freedom from utility, its delight in possibility for its own sake, its willingness to experiment and explore without rigid predetermined outcomes. When we creal, we participate in this divine play, this cosmic game of endless becoming. Heraclitus was also aware that war is omnipresent, the father of all things. Life is both traumatic and creative, but ultimately, trauma is weaker than the creative impetus. The wounds of trauma, understood as all that feels negative about life, are sometimes the resentment and agony of creatures against the tutelage of ongoing creation. The Creal moves not only by generating but also by decomposing, which may feel like a destruction.

Sometimes the best remedy to agony is sleep. Think about possibles in a dream: they are volatile, resistant to capture. Like cats, dreams are hard to catch. How do we produce creal states that will adhere to the real? Which dream fragments stay with us upon waking? Compare this to ideas for a book or a new way of living or dealing with our relationships: which ones will lead us all the way down a road of achievement? Crealing is daydreaming, swimming along the current of a generative dialectics between exploring possibility and finding what resonates deeply enough to persist and develop.

In crealing, we tune into these natural tendencies of creative emergence, neither forcing structure prematurely nor abandoning ourselves to mere chaos. What traditional Chinese philosophy calls wu-wei, "non-doing," is not passivity but action that aligns with the inherent tendencies of reality itself. Psychologist Mihaly Csikszentmihalyi's (1990) famous concept of "flow"—that state of absorbed creativity where self-consciousness dissolves, time seems to stand still and the entire universe be here and now—captures something essential about the experience of crealing. In flow, we're neither controlling nor controlled, but participating in a process that seems to move through us while we are stearing the boat without excessive effort. This reflects what philosopher Martin Buber (1970) called the "I-Thou" relationship with reality, where we relate not to objects but to presence, not to utility but to intrinsic value. The Creal is the Other in us, our cosmic twin.

Innovative transformation is often requested to show quantitative results, measurable outcomes. But excellent leaders know that qualitative features are more important, even if they are difficult to argue for in bureaucratic organizations. What is the relationship between crealing and quality? The novelist Robert Pirsig argued in his popular books *Zen and the Art of Motorcycle Maintenance* (1974) and *Lila* (1991) that Quality (with a capital Q) is in fact the primary reality, preceding both subjects and objects. Quality is not something we project onto reality but something we recognize within it, a fundamental aspect of Being itself. Quality is the meaning emerging from the fact that the Creal not only possibilizes diversity but also compossibilizes unity, as we well see in further chapters.

The relationship to meaning and truth in the Creal is radically different from conventional epistemology. We are used to thinking of truth as a set of verified stances, like scientific laws. People fear the idea that ultimate truth might be ultimate possibility, because that might mean the very negation of possibility itself by dogma. They associate truth with measurable effectiveness, with pragmatic utility and tangible facts. But truth in the Creal is inter-universal; it works across multiple universes and therefore may sometimes appear weak or counterfactual in any given world with a given set of laws. Truth is so general that it cannot be grasped in its entirety within the limits of a particular world. This does not mean that truth does not exist, but that we need a philosophy applicable across many worlds rather than the limited world described by current earthly science, with its morbid fascination for statistics and quantitative reductionism. Measuring is sometimes welcome and part of the equation, but so are the radically different unknowns. Crealing calls for this expanded, felt but shared truth, a planetary cosmology of the obvious and the hidden, a diplomatic dialogue about the seen and the unseen (Deleuze, 1994).

In crealing, we connect with this primordial Quality without form, the basis of all value. If you can create a new quality via innovation, a radically new experience or meaning, then you will disrupt the status quo. The Creal allows us—once in a very while—to invent new qualities, new values, new modes of significance that transcend existing categories and evaluations. This is what philosopher Nietzsche famously called transvaluation.

In wise souls, the will to quality is more radical than the will to power. Quality comes from the very nature of the Creal itself, from its tendency toward both differentiation and integration, both novelty and coherence. In crealing, we connect with our dual energy of dissemination and assemblage, becoming multiple and becoming one at the same time. It's a surrendering to a collective becoming, in which individual interest is not abandoned but paused or amplified.

As philosopher James P. Carse (1986) distinguishes in his work "Finite and Infinite Games," much of our conventional thinking operates within "finite games" aimed at winning, at terminating play with a definitive outcome. By contrast, crealing participates in the "infinite game," which aims not at winning but at continuing to play, not at closure but at opening. The infinite player, Carse writes, plays with boundaries, transforming them from barriers into horizons, from walls into doorways. This infinite play characterizes the Creal itself, which never exhausts its potential in any particular actualization.

Access and Obstacles: Who Can Creal?

What happens if someone cannot feel the Creal, cannot creal? That may be because the person is overly focused on *realing*, making things real, ruminating, naturalizing conventions as necessities, addressing problems to quickly fix, connecting machines and objects to protocols rather than diving into the ocean of potential. But there's promise even in seeing what we may call the "impotential," the dry and hopeless limitations of convicted realists to imagine beyond what they believe they know. The impotential can itself become a doorway to crealing if we question it carefully. The boundaries of the known may delineate the frontier of the unknown.

It would be a mistake to assume that only artists or traditionally "creative" people can effectively creal. I've found that even technically oriented minds, for instance engineers, often have a capacity for crealing once they give themselves permission. Their trained abilities to visualize complex systems and manipulate abstract models can become powerful assets in the crealing process, allowing them to navigate the ocean of potential with some dexterity, especially if they are working with natural forces like sources of energy. Environmental strategist Paul Hawken's comprehensive mapping of regenerative solutions reveals what he calls "the extraordinary healing capacity of nature" when human systems align with, rather than oppose natural processes (Hawken, 2021). Engineers, doctors and other technical practitioners notice more often than they might admit the inherent tendency of life systems to heal, grow, and fix themselves when conditions permit.

The main obstacles to crealing are not always intellectual limitations but emotional and social constraints: fear of the unknown, apparent security of the *status quo*, attachment to established identities and certainties, concern about others' judgments, horizontal rather than vertical thinking, domestic worries. Creating a safe environment where these constraints can temporarily relax is difficult but essential. Psychological safety, the shared belief that one won't be punished or humiliated for speaking up with ideas, questions, or concerns, is once again crucial for organizational crealing.

Some people cannot imagine anything beyond themselves, beyond their own interest, without some philosophically healthy training. Some cannot imagine a higher purpose or a world that is qualitatively better, perhaps because of too much egocentric nihilism or because they have been rewarded by luck all too often, and believe they are more competent than they really are. A lack of vision underlies their worldview even when they profess humanism and ethical concern. "One may smile and smile and be a villain," said Shakespeare via Hamlet. Crealing carries ethical implications that are philosophical before they can be made regulatory, an ethics of the self (daydreaming thoughtfully is a vital right for instance) and of the other (we may or may not share the same dreams). If we recognize ourselves as participants in reality's creative unfolding, we must acknowledge our responsibility for what we help bring forth. Not everything that can be created should be created. Not every possibility deserves actualization. The Creal itself tends toward what Leibniz called compossibility (see Chap. 5), the harmonious coexistence of diverse possibilities within a coherent world, rather than mere proliferation without integration.

Bohm (1980) spoke of the need for "proprioception of thought," the ability to observe our thinking processes in action, to catch ourselves when we fall into fragmentation or incoherence. Similarly, in crealing, we need awareness of the qualities and tendencies we're amplifying. Are we contributing to harmony or discord, to liberation or domination, to healing or harm? While a holistic answer may prove difficult, this ethical dimension connects crealing with the later phases of the crealectic method, and it is not impossible to assert if a given decision respects the ideal of a eudynamic balance between person, earth, work and society. In time, we may even be able to combine quantitative and qualitative approaches to eudynamia. And even in the crealing phase itself, we can cultivate awareness of the values implicit in our creative orientation. Because our creative acts will ripple through the web of reality, influencing countless other processes and possibilities, the combination of the SMILE_PH method (Sense-Making Interviews Looking at Elements of Philosophical Health) and the crealectic method should give us some clarity and reduce existential risks.

In crealing, our communion with the web of becoming precedes any innovation. We are called to create in ways that enhance rather than diminish the creative potential of the whole, primarily, and as much as possible of the living habitants of that whole, and such aspiration is found within us as we innernovate, diving into ourselves as creatures of creation.

From Crealing to Profusing

In the crealectic method, each phase creates the conditions for the next, in a natural unfolding that mirrors the creative process of reality itself. As we engage in crealing—reconnecting with the creative flux of Creal—we naturally begin to experience the first signs of a more or less elaborated upsurge of ideas, images, and possibilities. This leads organically to the next phase of the crealectic method: profusing, the free and abundant generation of possibilities without immediate editing or censorship, albeit within a given theme or encompassing direction.

The transition from crealing to profusing is like the shift from sensing the music to starting to dance. Where crealing is primarily receptive, though not passively so, profusing is more explicitly generative in a purposeful manner. If crealing reconnects us with the source, profusing allows that source to flow through us in specific expressions and explorations.

This flow should not be forced but will follow naturally from deep reconnection with the Creal. When we truly feel the creative pulse of reality, freely associated ideas and possibilities begin to bubble up without effort, like springs emerging from saturated ground. Our task in the upcoming profusing phase will be to allow this natural abundance in different thematic context, to suspend our habitual filtering and controlling tendencies, to let the creative flow express itself through us in all its exuberance, with only an ideal for frame. Csikszentmihalyi's extensive research on optimal experience reveals that creative flow emerges when challenge and skill are balanced, when self-consciousness dissolves, and when action and awareness merge (Csikszentmihalyi, 1990). What the crealing and profusing phases share experientially is a state of absorbed participation in creative becoming that transcends the subject-object dualism of ordinary consciousness: we create and are created at the same time.

But profusing without first crealing would be like trying to draw water from a well we haven't yet found. The quality and depth of our profusing depends directly on how fully and authentically we've reconnected with the natural psychedelism and abundance of Creal (drugs not needed here, as demonstrated by our capacity to dream).

Five Exercises to Practice Crealing

Like any profound shift in consciousness, crealing cannot be reduced to mechanical techniques. Nevertheless, certain practices can help cultivate the conditions for this reconnection with creative flow.

First, we must recognize that crealing is not something we do occasionally but something we are always engaged in, sometimes consciously or often not. We are always in the Creal, even when we don't recognize it. Our task is not to create crealing *ex nihilo* but to become aware of our ongoing participation in the creative pulse of reality. This is similar to how meditation practices in various traditions don't create consciousness but rather remove the obstacles to its natural presence.

In my workshops at Vattenfall R&D and elsewhere, but also in individual crealectic sessions, I've experimented with several approaches to facilitating crealing:

1. Evolutionary Visualization: Guiding participants through a visualization of evolutionary regression, moving backward down the tree of life, from humans to primates to earlier mammals, to reptiles, fish, and eventually single-celled organisms and the origin of life itself. This practice helps people feel their connection to the continuous creative process that has shaped all life forms, recognizing that we carry within us the creative history of the entire biosphere, a contiunuum of felt duration and metamorphic élan vital.

2. Natural Meditation: Standing near the Älvkarleby dam with Vattenfall engineers, I invited them to close their eyes and listen to the water rushing, imagining that its massive sound was the voice of life itself, the sound of Creal. The continuous flow of water serves as a perfect metaphor and embodiment of the continuous flow of creative potential. Other places and metaphorical sounds can be used, for instance a shimmering birch tree, with its unique leaves that flutter like a constellation of whispers in the slightest breeze.

3. Collective Randomness: A digital or analog bowl is filled with one hundred words that may or may not relate to the topic of the workshop. Each participant picks three words randomly and assembles them in the form of a sentence or verse related to the theme of the workshop, then a "multiverse" collective poem is made out of randomly sequencing each individual verse. This collective sense-making practice helps break down the barriers between individual imaginations, creating a shared field of creative potential, and a sense of trust in what the Creal may offer when we loosen the grip on trying to control the narrative.

4. Time-Lapse Perception: Showing a time-lapse video of natural processes—plants growing, clouds forming, landscapes changing, constructions being assembled by animals—helps shift perception from static objects to eudynamic processes. I ask participants to imagine that they could perceive reality at this accelerated pace, seeing the constant metamorphosis in everything around them.

5. Musical Immersion: Music directly accesses the preconceptual, flowing quality of experience. In some collective workshops or individual sessions, I've used carefully selected music to help participants bypass analytical thinking and connect directly with the felt sense of creative flow. Several musical pieces are effective at evoking the manifold quality of the Creal. For instance, and the list is purely indicative here as music can be a deeply subjective experience:

1. Arvo Pärt – *Fratres*
2. Kaija Saariaho – *Lichtbogen*
3. György Ligeti – *Atmosphères*
4. Alice Coltrane – *Journey in Satchidananda*
5. Anthony Braxton – *Composition No. 76*
6. Brian Eno – *An Ending (Ascent)*
7. Tim Hecker – *Virgins*
8. Ryoji Ikeda – *data.matrix*

9. Hildegard von Bingen – *Spiritus Sanctus Vivificans Vita*
10. Steve Reich – *Music for 18 Musicians*
11. Meredith Monk – *Dolmen Music*

These practices—examples among others—work best when approached not as techniques to master but as invitations to a different quality of awareness. The goal is not to force an experience but to create conditions where natural reconnection can occur. Ideally, we may invent the rules of the game in co-creation with the participant(s).

It is time to conclude this chapter. We began by standing at the shore where ocean meets land. We then found ourselves immersed in the waters of cosmology, feeling the rhythmic music of both metaphysical and physical waves within and around us. Crealing is this philosophical and emotional immersion, this felt participation in the creative flow that constitutes reality itself. It is not an esoteric practice for special occasions but a birthright we can reclaim in every moment, a natural mode of being that our fragmented utilitarian or hyper-realistic consciousness has temporarily obscured.

The Creative Real is not separate from us but is the very substance of our being and becoming. We are made of stars, as Carl Sagan reminded us, but those stars themselves are expressions of the cosmic creativity that continues to unfold in every instant in the vastness of the horizon that sews the known with the unknown. When we creal, we recognize and honor this fundamental truth of our being creatures. We remember what we've always known but learned to forget: that we are not isolated nor fixed subjects confronting a separated world, but participatory expressions of a universe that dreams itself into being, through us and others.

This remembering is not merely intellectual but embodied, felt, lived. It manifests as gratitude, wonder, joy, sometimes fear or anxiety, but more often a profound sense of homecoming. When we creal, we feel ourselves both expanded beyond our usual boundaries and more deeply connected to our essential nature, one that is large and contains multitudes, as per Walt Whitman's famous poem, *Song of Myself*. This paradoxical experience, becoming more fully ourselves by transcending ourselves into continuous creation, is the signature of authentic gratitude for the overabundance of life, beyond the trauma of life and death.

As we move forward into the profusing phase, we carry with us this felt sense of creative participation. We shall remember that our ideas and expressions are not merely personal productions but collaborations with the intercreative flow of reality itself. Every work is teamwork. We may approach this collaboration with both the humility it deserves and the dignity it bestows upon us as conscious participants in the great unfolding. We are not mere spectators; we have agency over our shared world: the next phases of the crealectic method (Profusing, Compossibilizing, and Realizing) will show how.

References

Barad, K. (2007). *Meeting the universe halfway: Quantum physics and the entanglement of matter and meaning*. Duke University Press.

Bergson, H. (2010 [1934]). *The creative mind: An introduction to metaphysics*. Dover Publications.

Berry, T. (2002). *The dream of the earth*. University of California Press.

Bloch, E. (1986). *The principle of hope*. MIT Press.

Bohm, D. (1980). *Wholeness and the implicate order*. Routledge.

Braidotti, R. (2013). *The posthuman*. Polity Press.

Buber, M. (1970 [1923]). *I and Thou*. Charles Scribner's Sons.

Carse, J. P. (1986). *Finite and infinite games*. Free Press.

Coole, D., & Frost, S. (Eds.). (2010). *New materialisms: Ontology, agency, and politics*. Duke University Press.

Csikszentmihalyi, M. (1990). *Flow: The psychology of optimal experience*. Harper & Row.

Damasio, A. (1999). *The feeling of what happens: Body and emotion in the making of consciousness*. Harcourt Brace.

de Miranda, L. (2017). On the concept of creal: The politico-ethical horizon of a creative absolute. In P. de Assis & P. Giudici (Eds.), *The dark precursor: Deleuze and artistic research* (pp. 510–516). Leuven University Press.

Deleuze, G. (1994). *Difference and repetition*. Columbia University Press.

Dietrich, A., & Kanso, R. (2010). A review of EEG, ERP, and neuroimaging studies of creativity and insight. *Psychological Bulletin, 136*(5), 822–848.

Fisher, M. (2009). *Capitalist realism: Is there no alternative?* Zero Books.

Gendlin, E. T. (1978). *Focusing*. Bantam Books.

Glăveanu, V. P., & de Miranda, L. (2021). Co-Creating the Real: A Transdisciplinary Dialogue. *Qualitative Inquiry, 28*(5), 476–485. https://doi.org/10.1177/10778004211063617 (Original work published 2022)

Haraway, D. (2016). *Staying with the trouble: Making kin in the Chthulucene*. Duke University Press.

Hawken, P. (2021). *Regeneration: Ending the climate crisis in one generation*. Penguin Books.

Heisenberg, W. (1999 [1958]). *Physics and philosophy: The revolution in modern science*. Prometheus Books.

Kauffman, S. (1993). *Origins of order: Self-organization and selection in evolution*. Oxford University Press.
Lakoff, G., & Johnson, M. (1999). *Philosophy in the flesh: The embodied mind and its challenge to Western thought*. Basic Books.
Leibniz, G. W. (1989 [1714]). *Philosophical essays*. Hackett Publishing.
Lovelock, J. (1979). *Gaia: A new look at life on earth*. Oxford University Press.
Maturana, H. R., & Varela, F. J. (1980). *Autopoiesis and cognition: The realization of the living*. D. Reidel Publishing Company.
Merleau-Ponty, M. (2012 [1945]). *Phenomenology of perception*. Routledge.
Penrose, R. (1989). *The emperor's new mind: Concerning computers, minds, and the laws of physics*. Oxford University Press.
Pirsig, R. M. (1974). *Zen and the art of motorcycle maintenance: An inquiry into values*. William Morrow & Company.
Pirsig, R. M. (1991). *Lila: An inquiry into morals*. Bantam Books.
Rilke, R. M. (2004). *Letters to a young poet*. W. W. Norton & Company.
Rudd, R. (2018). *The art of contemplation: Gentle paths to wholeness and self-realization*. Watkins Publishing.
Schrödinger, E. (1992 [1944]). *What is life?* Cambridge University Press.
Smolin, L. (2013). *Time reborn: From the crisis in physics to the future of the universe*. Houghton Mifflin Harcourt.
Spinoza, B. (1996). *Ethics*. Penguin Classics.
Varela, F. J., Thompson, E., & Rosch, E. (1991). *The embodied mind: Cognitive science and human experience*. MIT Press.
Whitehead, A. N. (1978 [1929]). *Process and reality: An essay in cosmology*. Free Press.

CHAPTER 4

Profusing

Abstract Profusing involves the uncensored generation of abundant possibilities without immediate judgment or filtering. Moving beyond traditional brainstorming, this chapter explores free association, divergent thinking, and the cultivation of cognitive abundance. It addresses the psychological and social obstacles to generative thinking while providing practical frameworks for individual and collective ideation that preserves diversity and encourages novel combinations across disciplines and domains.

Keywords Profusing • Divergent thinking • Free association • Brainstorming • Ideation • Cognitive abundance • Psychological safety • Creative freedom • Interdisciplinary thinking • Generative process

L. de Miranda, *Crealectics as a Creative Method*,
Palgrave Advances in Possibility Studies,
https://doi.org/10.1007/978-3-032-00427-7_4

Fig. 4.1 The profusing phase symbol

We have reset our minds, reconnected with the creative flow of life, and now stand at the threshold or volcano crater of manifested abundance. The first two phases of the crealectic method described in previous chapters were about making ourselves available to possibility by clearing mental space (resetting) and reconnecting with the creative flow that underlies all reality (crealing). We are only now ready to begin actively participating in this creative flow through the third phase: *Profusing* (Fig. 4.1).

Profusing embodies what we might in certain cases call "explosive ideation"—not always the steady accumulation of ideas but sometimes their sudden and disparate eruption. Like popcorn kernels reaching critical temperature, thoughts explode into consciousness with surprising intensity and unpredictable timing, and yet in other cases, they appear playfully like fish in the middle of corals.

The Exuberant Outpouring of Possibility

"Profusing" evokes a vivid imagery of ideas not just emerging but pouring forth and blooming disparately, ideas emerging forth in cascades of creative magma, each thought triggering others in chain reactions of possibility, ideally free from the constraints of censorship or imposed hierarchies. The term combines the notions of "profusion" (abundance, copious production) and "fusing" (merging, combining), suggesting both quantity and connection. It is a celebration of free association and unbridled creativity, where thoughts and inspirations explode, mingle, and flow with uninhibited vigor and playfulness (Csikszentmihalyi, 1990).

This phase represents the subtle move from receptive to generative engagement with the Creative Real. Where crealing reconnected us with the source of creative potential, profusing allows us to participate more actively in this creative process, generating a multitude of possibilities without immediate judgment or filtering. The philosopher Friedrich Nietzsche (1883–1885/2006) described in *Thus Spoke Zarathustra* the three metamorphoses of the spirit, from camel (carrying the burden of convention) to lion (destroying old values) to child (creating new values through play). Profusing embodies this third, childlike state where playful creation flows naturally from a spirit unburdened by conventional constraints. Contrary to most current brainstorming or ideation practices, profusing should not be approached without the "warming up" of the previous resetting and crealing phases.

The poet Walt Whitman (2001) proclaims in "Song of Myself": "I am large, I contain multitudes." Profusing embraces this multiplicity within us, allowing the diverse potentials within our consciousness to express themselves. This multitudinous nature reflects what philosopher William James (1890/1950) called the "stream of consciousness," the continuous flow of thoughts, sensations, and associations that constitutes our mental life. In profusing, we don't dam this stream but rather welcome it, *become it*, allowing ourselves to be carried by its current while remaining conscious of its movements. Curiosity researcher Scott Shigeoka's investigation into what he calls "deep curiosity" reveals that genuine inquiry requires what he terms "curiosity without agenda," the willingness to explore without predetermined outcomes (Shigeoka, 2024). This principle is essential for effective profusing, as attachment to specific results can unconsciously constrain the range of possibilities we're willing to consider or express. There are not many occasions in which we can practice "anything goes" in our societies: in profusing we get close to it, albeit within given themes.

Neurological studies have shown that the default mode network plays a crucial role in creative thinking and "aha" moments through its enhanced connectivity with cognitive control regions in highly creative individuals, supporting the generation of novel ideas and associations (Beaty et al., 2014). This network, active when our minds wander or daydream, plays a pivotal role in imaginating. When we let ideas fuse and pour out without self-imposed constraints, we're making space for this neural treasure trove to dance, allowing for richer, more diverse ideation. Neuroscientist Marcus Raichle, who pioneered research on the default mode network, describes it as a "symphony orchestra" of brain regions that coordinates activity even when we're not focused on specific tasks—precisely the kind of integrative, non-directed mental state that profusing cultivates.

The profusing phase bears resemblance to what psychologists call divergent thinking—the ability to generate multiple ideas or solutions to a single problem (Guilford, 1967). Unlike convergent thinking, which seeks a single "correct" answer, divergent thinking expands outward, exploring many possible or nearly impossible pathways. In the crealectic method, profusing is intentionally positioned before any convergent processes (which will come later in the compossibilizing phase) to ensure that innovative potential isn't prematurely limited. This sequencing is consistent with Teresa Amabile (1996), who emphasized that creative processes benefit from periods of open exploration before evaluative constraints are applied, as premature limitation of possibilities can hinder innovative outcomes.

Free Association and the Liberation of Mind

Profusing draws inspiration from the psychological technique of free association, a method pioneered by Sigmund Freud for psychoanalysis but which has applications far beyond therapy. In free association, one expresses thoughts without censorship or organization, allowing the mind to follow its natural pathways of connection. This technique reveals what William James (1890/1950) called the "associative threads" that connect our thoughts—connections that logical thinking often obscures but that can lead to unexpected insights.

Freud himself cited a letter from Friedrich Schiller about creative thinking, which maintained that "where there is a creative mind, Reason relaxes its watch upon the gates, and the ideas rush in pell-mell" (Freud, 1953). Similarly, the writer Ludwig Börne suggested to the young Freund that to foster creativity, "you write down, without any falsification or hypocrisy, everything that comes into your head" (Freud, 1953). These observations anticipated what modern creativity researchers call "cognitive disinhibition"—a reduced filtering of information that allows more unusual associations to enter consciousness (Carson, 2011).

Artist and creativity coach Julia Cameron says she practices what she calls "morning pages"—three pages of longhand, stream-of-consciousness writing done immediately upon waking, as a way to bypass the internal critic and access what she terms the "creative recovery" process (Cameron, 1992). This practice exemplifies the profusing principle of generating material without immediate evaluation, creating what Cameron calls a "spiritual chiropractic" that realigns our relationship with creative expression and life as creative process.

The concept of free association connects profusing to a broader philosophical tradition that questions the primacy of rational, linear thinking. Remember that philosophers like Henri Bergson (2007) distinguish between "intellect," which dissects reality into useful parts, and "intuition," which grasps the flowing continuity of experience. Bergson argued that intellect alone, with its focus on control, cannot fully apprehend the creative evolution of life; intuition is necessary to understand reality as process rather than protocolar state. Profusing cultivates this intuitive mode of cognition, momentarily suspending analytical dissection to experience disparate emergence.

Insight solutions involve activation of neural networks that process information more broadly and coarsely, allowing for connections across distantly related concepts that might not emerge through focused, analytical processing (Jung-Beeman et al., 2004). Research suggests that breakthrough creative ideas often occur when our brain's inner critic—located in the prefrontal cortex—temporarily steps back, giving space for unusual associations to surface (Dietrich, 2004). This neurological state resembles what jazz musicians describe when improvising—a flow state where conscious control recedes and creative expression emerges more spontaneously (Limb & Braun, 2008).

Psychologist Seymour Epstein (1994) proposed that humans operate with two parallel processing systems: a rational system that is analytical, logical, and relatively slow; and an experiential system that is holistic, associative, and rapid. While conventional problem-solving often prioritizes the rational system, creative breakthroughs frequently involve engaging the experiential system through techniques like free association. Profusing deliberately activates this experiential system, allowing it to generate possibilities that the rational system might never consider.

The Surrealist art movement embraced this approach through techniques like automatic writing and drawing, where artists attempted to bypass conscious control to access deeper creative resources. As André

Breton (1969) recommended in the *Surrealist Manifesto*: "Dictation of thought, in the absence of any control exercised by reason, outside of all aesthetic and moral preoccupation." While the crealectic method doesn't seek to eliminate reason, it does recognize the value of temporarily suspending critical judgment and analytic thinking to allow for greater creative expansion. The Surrealists' techniques produced not just art but also insights into the nature of creativity itself, demonstrating how conscious intent and unconscious process could collaborate rather than compete.

In their artistic experiments, the Surrealists discovered what psychologist Arthur Koestler (1964) would later formalize as "bisociation"—the creative connection of previously unrelated matrices of thought. Salvador Dalí's humorous melting clocks in "The Persistence of Memory" *bisociate* the rigid precision of timepieces with the fluidity of biological decay or experiential leisure, creating a new conceptual space that comments on both time and existentialism. Similarly, profusing encourages bisociative thinking by allowing diverse ideas to coexist and potentially connect in surprising ways. These possibles might not be compossible outside of our mind, but one must leave this evaluation for later stages.

In the corporate setting of Vattenfall R&D and similar organizations in which I applied the crealectic method, profusing invites participants to temporarily set aside their routine or logical mindsets and allow to generate ideas without immediate concern for practicality, feasibility, or consistency. This temporary suspension of judgment can be challenging for professionals trained in analytic intelligence, such as engineers, but it's precisely this challenge that makes the profusing phase valuable. It creates a space where conventional boundaries can be crossed and new connections formed. As one participant remarked after a workshop: "I spent years learning to filter out impossible ideas. Learning to temporarily disable that filter feels like rediscovering a muscle I forgot I had."

Beyond Brainstorming: The Art of Profusion

While profusing may resemble the familiar practice of brainstorming, it differs in subtle but important ways. Traditional brainstorming, as developed by advertising executive Alex Osborn in the 1950s, aims to generate solutions to specific problems through group ideation with deferred judgment (Osborn, 1953). Profusing, while embracing deferred judgment, is not necessarily problem-focused and extends beyond mere solution-finding to encompass broader thematic exploration. It shares more with what design thinking practitioners call "blue-sky thinking"—ideation that intentionally transcends current limitations and constraints (Brown, 2009). The crealectic method is intentionally idealistic.

Research into the effectiveness of brainstorming has revealed some limitations. Creativity loss can occur during group brainstorming due to factors like evaluation apprehension (fear of judgment despite rules against criticism), production blocking (inability to express ideas while others are speaking or even listening), and "free riding" (reduced individual effort in groups) (Diehl & Stroebe, 1987; Mullen et al., 1991). These findings challenge the popular assumption that groups naturally enhance creativity, suggesting instead that group processes can sometimes inhibit rather than facilitate ideation.

Social psychologist Adrian Furnham (2000) has argued that traditional brainstorming suffers from "collaborative fixation"—the tendency for group members to fixate on similar ideas, reducing the diversity of perspectives. In one revealing study, groups generated fewer ideas than the same number of individuals working alone, despite participants' subjective impression that the group session was more productive (Nijstad et al., 2006). This illustrates what psychologists call the "illusion of group productivity"—we tend to feel more creative in groups even when objective measures indicate otherwise.

Addressing these limitations, studies have shown that individual ideation followed by group discussion often produces better results than pure group brainstorming (Paulus & Nijstad, 2003). In general, in my interventions in corporations or institutions, I combine individual sessions of dialogue with collective workshops. This sequencing harnesses both the divergent thinking advantages of individual work and the elaborative benefits of group interaction. Research by organizational psychologist Leigh Thompson shows that combining individual ideation with structured team interaction generates more creative outcomes than traditional group brainstorming, and includes techniques for neutralizing alpha-dominant behaviors that can inhibit group creativity.

In crealectic sessions, we may incorporate these research insights by structuring profusing exercises in ways that mitigate potential limitations. For example, in Vattenfall workshops, participants were often asked to profuse via individual written ideation before sharing and building upon each other's ideas. This approach, sometimes called "brainwriting," helps ensure that dominant personalities don't overshadow quieter contributors and that everyone can generate ideas without interference (Heslin, 2009). Electronic variations of brainwriting using digital tools can further reduce social inhibitions by allowing anonymous contributions, particularly valuable in organizations with strong hierarchical structures or in cultures where direct challenge of authority figures is discouraged (Dennis & Williams, 2003).

Computer scientists Kenneth Stanley and Joel Lehman's research on evolutionary algorithms reveals a counterintuitive principle: the most significant innovations often emerge not from direct goal-seeking but from what they call "novelty search," the deliberate pursuit of behavioral novelty and interesting differences without predetermined objectives (Stanley & Lehman, 2015). This research provides computational support for the profusing principle that abundance and diversity of possibilities often matter more than immediate practicality or goal-directedness.

Social psychologist Richard Hackman (2011), in his extensive research on effective teams, identified the importance of what he called "enabling structures"—frameworks that simultaneously provide direction and freedom. The crealectic method implements such enabling structures through clear semi-structured processes that paradoxically enhance freedom by removing the cognitive load of deciding how to proceed. Participants can focus their mental energy entirely on generative thinking and the theme of the workshop or session rather than procedural concerns.

The metaphor of brainstorming—reminiscent perhaps of the Romantic *Sturm und Drang*—suggests a violent weather event, sudden and intense but limited in duration if not repeated. Profusing, by contrast, evokes a wellspring rather. The crealectic method is ideally not about one-shot ideation but about creating a regular space for the continuous and gentle emergence or pruning of possibilities. This aligns with what creativity researcher Keith Sawyer (2012), inspired by Deleuze, calls "zig-zag creativity"—the recognition that creative processes often proceed through iterative cycles, requiring sustained engagement rather than occasional bursts of activity.

Profusing may embrace ambiguity as a productive state rather than a problem to be resolved. Building on psychologist Todd Lubart's (2001) research demonstrating correlations between tolerance for ambiguity and creative achievement across domains, the comfort with unresolved possibilities that profusing cultivates may itself enhance creative capacity. As one workshop participant remarked: "Usually when we brainstorm, I'm already mentally evaluating each idea as it comes. This profusing approach feels different—I'm allowed to just let the ideas flow without immediately judging them. It's liberating." This liberation from premature evaluation—easier said than done indeed—aligns with what Csikszentmihalyi (1996) identifies as a key characteristic of highly creative individuals: the ability to delay closure and remain open to emerging possibilities longer than others might find comfortable.

Abundance is qualitative. Economists Marian Tupy and Gale Pooley's analysis of resource availability reveals a counterintuitive principle: genuine scarcity is primarily a function of limited creativity rather than finite resources (Tupy & Pooley, 2022). Their research shows that human ingenuity consistently transforms apparent limitations into opportunities for innovation, supporting the profusing principle that abundance of ideas creates abundance of solutions.

Freedom, Framework, and Fear

While profusing celebrates creative freedom, it's important to insist that this freedom exists within a framework. Complete randomness rarely produces meaningful innovation; rather, creativity thrives when it has some methodical or thematic constraint to push against (Stokes, 2005). This relationship between freedom and constraint has been observed across creative domains. Jazz improvisation, for instance, offers musicians tremendous freedom but within the structure of chord progressions and musical scales. Similarly, poets might embrace free verse while still working within the thematic or emotional constraints of their subject matter. As Stravinsky famously noted, perhaps with some masochistic exaggeration, "The more constraints one imposes, the more one frees oneself" (Stravinsky, 1942). Moreover, for emergence to occur, we need a multiplicity of constituting elements. For the latter stage of compossibility to be robust (see Chap. 5), we need a pell-mell of possibilities, sometimes bordering contradiction, as in disparate worlds coming into contact.

Usually, a collective crealectic workshop will have a theme that functions like a space of possibility. In our work with Vattenfall R&D, the framework for profusing might have been for instance "holistic energy systems" or "regenerative consciousness in 2030." These themes must be broad enough to allow for wide-ranging ideation but focused enough to give the process direction and relevance, a horizon. Too much freedom may kill the desire for action. Without such framing, profusing risks becoming too disconnected from the practical concerns that may have initiated the creative process. More importantly, it's about looking in the same direction, albeit from various angles and perspectives, to create a sense of mindful esprit de corps.

A given theme serves as what complexity theorists might call a "strange attractor"—a pattern within apparent chaos that guides the system's behavior without rigidly determining it (Gleick, 1987). In profusing, the theme of the workshop acts as this attractor, creating coherence amid diversity without imposing uniformity. While collective workshops in philosophical health and crealectics need a theme to federate the participants, in individual sessions, the implicit theme might simply be the person's life.

Despite its apparent simplicity, profusing may face numerous psychological and social obstacles. Perhaps the most significant is fear; fear of judgment, fear of appearing foolish by being too personal, fear of proposing ideas that might later be rejected. This fear is not irrational—in many organizational cultures, proposing unconventional or personal ideas can indeed carry social stigma and professional risks.

The process also confronts our deeply ingrained habits of self-censorship. We routinely filter our thoughts before expressing them, a skill that serves us well in many contexts but can severely limit creative exploration. As Edward de Bono (1970) noted in his work on lateral thinking, our brains naturally prefer well-worn neural pathways, making truly novel thinking uncomfortable and sometimes even physically demanding.

Cultural conventions represent another obstacle. What constitutes an "appropriate" idea varies tremendously across cultures, genders, or organizations. In some environments, proposing radical change may be encouraged; in others, it might be seen as challenging authority or disrupting group harmony. The crealectic facilitator must be sensitive to these diversity dimensions, creating an equal space where participants feel safe to venture beyond convention while respecting essential organizational values.

Corporate cultures often inadvertently discourage profusing through implicit norms that privilege practicality over possibility, certainty over exploration, realism over idealism, and immediate measurable results over developmental processes. As Boltanski and Chiapello (2005) observe in *The New Spirit of Capitalism*, modern organizations may incorporate creative exercises as temporary diversions that ultimately reinforce rather than challenge existing structures. To avoid this trap, the crealectic method insists that compossible thinking is not merely a creative interlude but an essential component of genuine innovation, one that ideally should be implemented for the long-term. Having one annual corporate gathering in which anything goes, completed perhaps by some fun physical challenges, is rather worthless in terms of deep innovation.

To address these obstacles, it is not enough to establish what Edmondson (1999) calls "psychological safety"—an environment where people believe they won't be punished or humiliated for speaking up with ideas, questions, or concerns. Moreover, groups should have the time to let novelty unfold and the crealectic culture to infuse over regular thematic meetings that are not too separated from the real processes of the company or institution.

Quality and Time

While profusing emphasizes quantity and diversity of ideas without immediate judgment, this doesn't mean that all ideas will be considered equally valuable.

The quality that matters in profusing is not the immediate practicality or feasibility of individual ideas but rather their generative potential—their ability to stimulate further thinking, reveal unexpected connections, or challenge limiting assumptions. Ideas that appear impractical or even absurd may contain seeds of insight that, when later combined with other elements in the compossibilizing phase, yield powerful innovations.

Interdisciplinary thinking is key. Truly innovative ideas often emerge at the intersection of disciplines or domains. Profusing creates opportunities for such intersections by encouraging the expression of diverse perspectives and allowing unusual combinations to form. The quality emerges not from individual ideas in isolation but from their interactions and the new possibilities they might collectively reveal. Practically speaking, it is recommended, where possible, that the participants in the workshop come from diverse expertises and aspects of work.

Participants might propose ideas that initially seem technically unfeasible or economically impractical. Rather than dismissing these as "low quality," we may recognize them as potentially valuable elements in the larger creative ecosystem. An apparently impractical or "foolish" idea like "meditation-enhanced conductivity" might later connect with more conventional approaches to energy efficiency in ways that wouldn't have emerged if the unusual or absurd idea had been filtered out prematurely.

This approach aligns with what innovation researchers call "loonshots"—seemingly crazy ideas that can ultimately lead to breakthrough innovations (Bahcall, 2019). The history of technological development is replete with examples of ideas initially dismissed as impractical or even absurd that later became transformative innovations. The profusing phase creates space for such loonshots to emerge and sometimes be preserved for potential development or articulation within more habitual protocols.

The relationship between profusing and time deserves special consideration. While analytics privileges efficiency and often seeks to compress creative processes into the shortest possible timeframe, profusing recognizes that creative emergence often requires a different temporal approach. Ideas need time to germinate, combine, and evolve—a process that cannot always be accelerated without sacrificing depth and originality.

Remember that French philosopher Bergson (2007) distinguished between "clock time" (temps) and "lived time" (durée), the former being homogeneous, divisible, and measurable; the latter being heterogeneous, continuous, and qualitative. Profusing operates primarily in the realm of duration, where the significance of a moment isn't measured by its minutes but by its creative intensity and its connection to the continuum of consciousness.

In practical terms, this means that profusing sessions in fact needn't be lengthy to be effective, especially because they have been prepared by moments of resetting and crealing. A fifteen-minute period of intensely focused ideation might generate more meaningful possibilities than hours of routine thinking. Conversely, some creative processes may benefit from extended incubation periods, where ideas develop below the threshold of conscious attention, hence the need for regular innovative meetings. The crealectic method accommodates both approaches, recognizing that different challenges and different participants may require different temporal frameworks.

One might structure a workshop series to include both intensive profusing sessions and periods of "creative incubation" where participants step away from direct engagement with the challenge. Research on incubation effects suggests that taking breaks from conscious problem-solving can facilitate creative insights when returning to the task (Sio & Ormerod, 2009). This rhythmic alternation between focused ideation and relaxed incubation often yields richer results than sustained effort alone.

The Ethics of Profusing: Power, Voice, and Diversity

Profusing has both individual and collective dimensions. While each participant generates their own ideas, these ideas emerge within a shared space and are influenced by the collective energy and direction of the group. This relationship between individual and collective creativity reflects philosopher Gilbert Simondon's (2020) concept of "transindividuality"—the idea that individuation (the process of becoming an individual) always occurs within and through collective processes.

In the crealectic method, we recognize that genuine innovation rarely occurs in isolation, although some personal isolation is likely to be needed at some points of the process. Even seemingly individual breakthroughs are typically embedded in networks of influence, conversation, and collaborative thinking. The profusing phase deliberately cultivates these networks, creating conditions where ideas can cross-pollinate and evolve through collective engagement.

This approach challenges the myth of the lone genius that pervades much of Western thinking about innovation. As Keith Sawyer (2007) demonstrates in his research on collaborative creativity, significant innovations typically emerge from a dialectic between group processes and individual inspiration. Profusing embraces this transindividual nature of creativity, structuring exercises to maximize productive interaction while still respecting individual contributions and the autonomy of the person. Institutions and corporations should allow individuals to cultivate their secret gardens via autonomous exploration, while making sure these are not considered as hobbies disconnected from the creative world and the purpose of the group. Innernovation should not be introversion, and spaces for shared ideation should also be cultivated: in 2023, Vattenfall R&D asked me to create a library of 300 books, the titles and themes of which were distributed along the 5 steps of the method. Having a diverse and interdisciplinary library at the heart of a building, allowing people to pause and read without feeling guilty, or have a comfortable coffee-break conversation surrounded by books is enough to plant a crealectic and philosophically healthy seed in the midst of routines.

In crealectic workshops, this creative dialectic between the person and the group (which is the theme of my doctoral work and book *Ensemblance*) might take the form, for instance, of "collaborative cascades," where participants begin with individual ideation, then share in pairs, then in quads, and finally as a whole group. This progressive expansion allows ideas to evolve through multiple interactions while ensuring that every voice contributes to the collective output. The resulting ideas are neither purely individual nor homogeneously collective but emerge from the dynamic interplay between personal and shared creativity.

The profusing phase, when performed collectively, raises important ethical questions about whose ideas are expressed, heard, and valued. In any group, power dynamics affect who feels comfortable speaking, whose contributions are taken seriously, and which perspectives shape the emerging direction. The crealectic method acknowledges these dynamics and works to create conditions where diverse voices can meaningfully contribute.

Research on group creativity consistently shows that intellectual or cultural diversity—of backgrounds, disciplines, perspectives, and thinking styles—enhances innovative potential (Page, 2007). Homogeneous or familiar groups may work together smoothly but tend to produce less creative outcomes than more cognitively diverse groups. The challenge lies in structuring interactions so that deep (not superficial) diversity becomes a creative asset rather than a source of conflict or confusion. Remember that diversity is about ways of thinking and acting, as well as positions in the hierarchy, not the way our bodies or clothes look.

We may employ several strategies to address power imbalances and maximize the benefits of diversity. Anonymous ideation techniques allow contributions to be evaluated on their merits rather than their source, while non-anonymous ideation, if repeated carefully, may encourage mental courage. Structured turn-taking ensures that quieter voices aren't overshadowed by more dominant ones. Deliberately heterogeneous groupings may bring together different disciplines and levels of seniority to create productive creative friction.

The profusing phase also confronts the ethical question of what kinds of ideas should be expressed. While we emphasize freedom from censorship to maximize creative potential, this doesn't mean eliminating all ethical boundaries. The distinction lies between suspending judgment about practicality or convention (which enhances creativity) and suspending judgment about fundamental ethical principles (which could lead to harmful outcomes). However, in some cases, ethical guidelines are too attentive to details and bureaucratic, and they hinder creative thinking and risk taking. Encouraging fragility does not always nurture resilience.

Rather than imposing rigid ethical guidelines that might constrain creative thinking, we cultivate deep listening or what philosopher Miranda Fricker (2007) calls "virtuous listening"—an approach that remains open to diverse perspectives while maintaining awareness of power dynamics and the higher purpose of the group. This allows the profusing phase to generate innovative possibilities while remaining grounded in a commitment to beneficial outcomes, along the care for eudynamia between person, earth, work and society.

Practical Approaches to Profusing

How do we translate these theoretical principles into practical workshop exercises? Based on our experience, several approaches have proven effective:

1. *Individual Written Profusion*: Participants silently write down as many ideas as possible related to the theme, without pausing to evaluate them. Research suggests that writing by hand may activate different neural pathways than typing, potentially enhancing creative thinking (Mueller & Oppenheimer, 2014). The physical act of writing can also slow down thinking just enough to capture ideas more completely, while keeping keyboard and digital routines aside.

2. *Round-Robin Sharing*: In small groups, participants take turns sharing one idea at a time, building on each other's contributions. This structured approach ensures everyone has equal opportunity to contribute and prevents dominant personalities from monopolizing the conversation. It also creates a rhythm that can help sustain the creative flow longer than unstructured discussion might.

3. *Reverse Thinking*: Participants intentionally generate "wrong" answers or opposite approaches to the challenge at hand. This technique, similar to what Edward de Bono called "provocations," deliberately disrupts conventional thinking patterns. By temporarily inverting goals or assumptions, we can discover novel perspectives that might not emerge through direct approaches.

4. *Visual Profusing*: Using drawing, mind mapping, or collage to express ideas visually rather than verbally. This approach engages different cognitive processes and can be especially valuable for expressing concepts that resist verbal articulation. Visual thinking activates spatial reasoning and pattern recognition in ways that complement verbal ideation.

5. *Physical Movement*: Incorporating movement into the profusing process. Research has shown that physical activity can enhance creative thinking (Oppezzo & Schwartz, 2014). We might have participants walk while sharing ideas or use physical space to organize emerging concepts. This embodied approach recognizes that creativity isn't just a mental process but engages our whole being.

The effectiveness of these approaches, which are examples among others, varies depending on the participants, the theme, and the specific organizational context. A skillful facilitator will adapt methods to the group's needs, perhaps starting with more structured approaches and gradually introducing greater freedom as participants become comfortable with the profusing mindset. It is for the convenor to be bold and not assume for instance that "engineers don't like poetry" or other clichés.

From Profusing to Compossibilizing: The Transition

As the profusing phase generates an abundance of possibilities, we begin to approach the natural transition to the next phase of the crealectic method: compossibilizing. This transition isn't abrupt but gradual, as participants begin to notice patterns, connections, and potential harmonies among the diverse ideas they've generated. This organic movement from divergence to emergence reflects what systems theorist Joanna Macy (2007) calls "the Great Turning"—a shift from fragmentation toward reconnection and integration.

The profusing phase might be compared to a garden with seeds of many different plants sprouting simultaneously. Not all will reach maturity, but each contributes to the overall ecosystem. The transition to compossibilizing is like the gardener beginning to see which plants might flourish together, which need more space or pruning, and which combinations might create a harmonious landscape. Ecological philosopher David Abram (1996) describes this as "participatory perception"—a way of seeing that recognizes patterns and relationships rather than isolated objects.

In practice, this transition often manifests as a shift in the quality of conversation. Where earlier discussions focused on generating and expressing ideas, participants now begin asking questions like "How might these approaches work together?" or "What patterns do we see emerging?" These questions may arise organically from the profusion itself, as the mind naturally seeks coherence amid diversity. Gestalt psychologists call this tendency "Prägnanz"—the mind's inclination to organize perceptions into meaningful wholes (Koffka, 1935). This natural organizing tendency serves as a bridge between the abundant multiplicity of profusing and the harmonious integration of compossibilizing.

The transition from profusing to compossibilizing also mirrors what Whitehead (1929/1978) called "concrescence," the process by which disparate elements come together to form new unities. For Whitehead, this wasn't a reduction of diversity but rather its integration into more complex wholes. Similarly, the movement from profusing to compossibilizing doesn't diminish the richness generated in the previous phase but reconfigures it into potentially viable patterns.

Complexity theorist Stuart Kauffman (1995) describes how complex systems naturally move between states of chaos and order, with the most creative and adaptive systems operating at "the edge of chaos"—a zone where there is enough stability for coherent patterns to form but enough fluidity for novel configurations to emerge. The transition from profusing to compossibilizing navigates this edge, allowing organization to emerge without imposing rigid structures that would eliminate creative potential.

Design theorist Herbert Simon (1996) noted that complex systems often exhibit "near-decomposability:" they can be understood as hierarchies of interconnected subsystems. The transition to compossibilizing often involves this kind of pattern recognition, where participants begin to see how certain clusters of ideas might form coherent subsystems within a larger whole. Information theorist and ecologist Gregory Bateson (1979) called this perception of pattern "the pattern that connects," suggesting that such recognition represents a higher order of learning than the accumulation of discrete facts or ideas.

We might facilitate this transition by asking participants to identify ideas they find particularly intriguing or promising, not to narrow possibilities prematurely but to begin recognizing potential centers of gravity around which compossible clusters might form. This initial clustering provides raw material for the more deliberate harmonization that will occur in the compossibilizing phase. Some believe that visual mapping techniques—physically arranging ideas in space to reveal relationships—can be particularly effective for knowledge management, engaging spatial cognition to complement verbal reasoning (Eppler & Burkhard, 2007).

One participant described this transition moment eloquently: "It's like when you're looking at stars in the night sky. At first, you just see individual points of light. Then suddenly, you start seeing constellations or figures, patterns that give meaning to what seemed random before." This perceptual shift illustrates what cognitive scientist Douglas Hofstadter (1979) calls "strange loops"—the emergence of patterns at one level of perception that couldn't be predicted from examining the elements in isolation. The CIPHER model (see the introduction of the present book) can function as such a loop.

This transition doesn't invalidate or abandon the creative abundance generated during profusing. Rather, it represents the next natural stage in the creative process, where the rich field of possibilities begins to organize into potential worlds, collections of compatible possibilities that might coexist coherently. The compossibilizing phase will continue this process more deliberately, building directly upon the foundation established through profusing. As environmental philosopher Arne Naess (1989) might express it, this movement represents the shift from "shallow" to "deep" ecology, from seeing isolated elements to perceiving their interconnections within living systems.

Conclusion: The Fertile Abundance

The profusing phase stands as a celebration of creative abundance, a deliberate cultivation of possibility before any concern with practicality or coherence. It embodies what philosopher Deleuze (1994) called "the affirmation of difference," embracing multiplicity and divergence as creative forces in themselves rather than problems to be resolved.

By temporarily suspending our habitual filtering mechanisms and allowing ideas to emerge in profusion, we created conditions for genuine and more personal innovation. The manifold quantity and diversity of possibilities generated during this phase provide the raw material from which truly novel approaches can later be crafted. Without this abundant foundation, later phases would have limited material to work with, potentially resulting in incremental rather than transformative innovation.

Yet profusing is not merely a means to an end, valuable only for the possibilities it generates for later development. The experience itself holds intrinsic value, reconnecting participants with their inner freedom and creative capacities, while challenging artificial boundaries between disciplines, domains, and ways of knowing. Many workshop participants report that the profusing phase changes not just what they think but how they think—or simply that they *think* and *imagine* rather than just calculate, worry, project or anticipate—opening new cognitive pathways that may be cultivated beyond the specific creative challenge.

Common parlance tends to use the term thinking for activities like ruminating, calculating, worrying, etc. We should reserve the term thinking to its philosophical, cosmological, spiritual, or compossible aspect. One of the core texts on the question is Heidegger's essay "What Is Called Thinking?" For Heidegger (1968) and many other philosophers, we are rarely thinking even though we live in an age obsessed with information, logic, and rational planning. True thinking is not calculation nor only problem-solving; it is receptive and creative openness to Being, Becoming or what we called the Creal, the Creative Real. What calls us to think and dream? Thinking is sometimes a way of responding to a calling, a dialogue with the cosmos.

As we stand amid this abundance of possibilities for thinking and world-building, we may begin to intuit or decipher patterns and potential harmonies that will guide us into the next phase of the crealectic method. The multiverse of ideas generated through profusing contains infinite potential futures. In the compossibilizing phase, we will begin to explore which of these futures might help to form coherent, harmonious worlds, not by eliminating diversity although some pruning will be needed but by discovering how diverse elements might coexist in eudynamic harmony.

References

Abram, D. (1996). *The spell of the sensuous: Perception and language in a more-than-human world*. Pantheon Books.

Amabile, T. M. (1996). *Creativity in context: Update to the social psychology of creativity*. Westview Press.

Bahcall, S. (2019). *Loonshots: How to nurture the crazy ideas that win wars, cure diseases, and transform industries*. St. Martin's Press.

Beaty, R. E., Benedek, M., Wilkins, R. W., Jauk, E., Fink, A., Silvia, P. J., et al. (2014). Creativity and the default network: A functional connectivity analysis of the creative brain at rest. *Neuropsychologia, 64*, 92–98.

Bateson, G. (1979). *Mind and nature: A necessary unity*. E. P.

Bergson, H. (2007). *Creative evolution* (A. Mitchell, Trans.). Palgrave Macmillan. (Original work published 1911).

Boltanski, L., & Chiapello, E. (2005). *The new spirit of capitalism*. Verso.

Brown, S. (2009). *Play: How it shapes the brain, opens the imagination, and invigorates the soul.* Avery.

Breton, A. (1969). *Manifestoes of surrealism* (R. Seaver & H. R. Lane, Trans.). University of Michigan Press.

Cameron, J. (1992). *The artist's way: A spiritual path to higher creativity.* Jeremy P. Tarcher/Putnam.

Csikszentmihalyi, M. (1990). *Flow: The psychology of optimal experience.* Harper & Row.

Carson, S. H. (2011). *Your creative brain: Seven steps to maximize imagination, productivity, and innovation in your life.* Jossey-Bass.

Csikszentmihalyi, M. (1996). *Creativity: Flow and the psychology of discovery and invention.* Harper Collins.

de Bono, E. (1970). *Lateral thinking: Creativity step by step.* Harper & Row.

Dennis, A. R., & Williams, M. L. (2003). Electronic brainstorming: Theory, research, and future directions. In P. B. Paulus & B. A. Nijstad (Eds.), *Group creativity: Innovation through collaboration* (pp. 160–178). Oxford University Press.

Deleuze, G. (1994). *Difference and repetition* (P. Patton, Trans.). Columbia University Press. (Original work published 1968).

Diehl, M., & Stroebe, W. (1987). Productivity loss in brainstorming groups: Toward the solution of a riddle. *Journal of Personality and Social Psychology, 53*(3), 497–509.

Dietrich, A. (2004). The cognitive neuroscience of creativity. *Psychonomic Bulletin & Review, 11*(6), 1011–1026.

Edmondson, A. (1999). Psychological safety and learning behavior in work teams. *Administrative Science Quarterly, 44*(2), 350–383.

Eppler, M. J., & Burkhard, R. A. (2007). Visual representations in knowledge management: Framework and cases. *Journal of Knowledge Management, 11*(4), 112–122.

Epstein, R. (1994). *Cognition, creativity, and behavior: Selected essays.* Praeger.

Freud, S. (1953). *The interpretation of dreams* (J. Strachey, Trans.). Hogarth Press. (Original work published 1900).

Fricker, M. (2007). *Epistemic injustice: Power and the ethics of knowing.* Oxford University Press.

Furnham, A. (2000). The brainstorming myth. *Business Strategy Review, 11*(4), 21–28.

Gleick, J. (1987). *Chaos: Making a new science.* Viking.

Guilford, J. P. (1967). *The nature of human intelligence.* McGraw-Hill.

Hackman, J. R. (2011). *Collaborative intelligence: Using teams to solve hard problems.* Berrett-Koehler Publishers.

Heidegger, M. (1968). *What is called Thinking?* (F. D. Wieck & J. G. Gray, Trans.). Harper & Row. (Original work published 1954).

Heslin, P. A. (2009). Better than brainstorming? Potential contextual boundary conditions to brainwriting for idea generation in organizations. *Journal of Occupational and Organizational Psychology, 82*(1), 129–145.

Hofstadter, D. R. (1979). *Gödel, An eternal golden braid*. Basic Books.

James, W. (1890/1950). *The principles of psychology*. Dover Publications.

Kauffman, S. A. (1995). *At home in the universe: The search for laws of self-organization and complexity*. Oxford University Press.

Jung-Beeman, M., Bowden, E. M., Haberman, J., Frymiare, J. L., Arambel-Liu, S., Greenblatt, R., et al. (2004). Neural activity when people solve verbal problems with insight. *PLoS Biology, 2*(4), e97.

Koffka, K. (1935). *Principles of Gestalt psychology*. Harcourt, Brace & World.

Koestler, A. (1964). *The act of creation*. Macmillan.

Limb, C. J., & Braun, A. R. (2008). Neural substrates of spontaneous musical performance: An fMRI study of jazz improvisation. *PLoS One, 3*(2), e1679.

Lubart, T. I. (2001). Models of the creative process: Past, present and future. *Creativity Research Journal, 13*(3–4), 295–308.

Macy, J. (2007). *World as lover, world as self: Courage for global healing*. Parallax Press.

Mueller, P. A., & Oppenheimer, D. M. (2014). The pen is mightier than the keyboard: Advantages of longhand over laptop note taking. *Psychological Science, 25*(6), 1159–1168.

Mullen, B., Johnson, C., & Salas, E. (1991). Productivity loss in brainstorming groups: A meta analytic integration. *Basic and Applied Social Psychology, 12*(1), 3–23.

Naess, A. (1989). *Ecology, community and lifestyle: Outline of an ecosophy*. Cambridge University Press.

Nietzsche, F. (1883–1885/2006). *Thus spoke Zarathustra: A book for all and none*. Cambridge University Press.

Nijstad, B. A., Stroebe, W., & Lodewijkx, H. F. M. (2006). The illusion of group productivity: A reduction of failures explanation. *European Journal of Social Psychology, 36*(1), 31–48.

Oppezzo, M., & Schwartz, D. L. (2014). Give your ideas some legs: The positive effect of walking on creative thinking. *Journal of Experimental Psychology: Learning, Memory, and Cognition, 40*(4), 1142–1152.

Osborn, A. F. (1953). *Applied imagination: Principles and procedures of creative thinking*. Scribner.

Page, S. E. (2007). *The difference: How the power of diversity creates better groups, firms, schools, and societies*. Princeton University Press.

Paulus, P. B., & Nijstad, B. A. (Eds.). (2003). *Group creativity: Innovation through collaboration*. Oxford University Press.

Sawyer, K. (2007). *Group genius: The creative power of collaboration*. Basic Books.

Sawyer, R. K. (2012). *Explaining creativity: The science of human innovation* (2nd ed.). Oxford University Press.

Simon, H. A. (1996). *The sciences of the artificial* (3rd ed.). MIT Press.

Shigeoka, S. (2024). *Seek: How curiosity can transform your life and change the world.* Harmony Books.

Simondon, G. (2020). *Individuation in light of notions of form and information* (T. Adkins, Trans.). University of Minnesota Press. (Original work published 1989).

Sio, U. N., & Ormerod, T. C. (2009). Does incubation enhance problem solving? A meta-analytic review. *Psychological Bulletin, 135*(1), 94–120.

Stanley, K. O., & Lehman, J. (2015). *Why greatness cannot be planned: The myth of the objective.* Springer.

Stokes, P. D. (2005). *Creativity from constraints: The psychology of breakthrough.* Springer.

Stravinsky, I. (1942). *Poetics of music in the form of six lessons.* Harvard University Press.

Tupy, M. L., & Pooley, G. L. (2022). *Superabundance: The story of population growth, innovation, and human flourishing on an infinitely bountiful planet.* Cato Institute.

Whitehead, A. N. (1929/1978). *Process and reality: An essay in cosmology.* Free Press.

Whitman, W. (2001). *Song of myself.* Dover Publications. (Original work published 1855).

CHAPTER 5

Compossibilizing

Abstract Compossibilizing draws on Leibniz's concept of compossibility to organize diverse possibilities into dynamically harmonious, coherent worlds. Rather than eliminating contradiction, this phase seeks eudynamic compatibility among different elements, creating what the chapter terms eudynamia between person, earth, work and societies. It explores practical approaches to pattern recognition, systems thinking, and the creation of coherent innovation clusters that preserve diversity while enabling functional integration and mutual enhancement.

Keywords Compossibilizing • Compossibility • Leibniz • Harmony • Systems thinking • Compatibility • Pattern recognition • Integration • Coherence • Symbiosis

L. de Miranda, *Crealectics as a Creative Method*,
Palgrave Advances in Possibility Studies,
https://doi.org/10.1007/978-3-032-00427-7_5

Fig. 5.1 The compossibilizing phase symbol

We have created a multiverse of possibility, a profusion of ideas, some of which may be suggesting a world of its own, that is, a set of compossible situations aligned closely enough that the realities can be part of the same coherent and healthy system. Now some of these imagined profused possibilities may not be compatible with one another. The task before us in this new step of the crealectic method is to group our profused ideas into clusters of compossibility, or compatibility (Fig. 5.1).

Compossibility: The Leibnizian Heritage

Once we've produced a diverse and rich set of disparate ideas, we must examine how some of them can coexist within the same desired world. "Compossibilizing" emerges as a holistic approach that seeks harmony amid a multitude of possibilities, maximizing their interconnection as much as possible, but no more than is compossible, that is, not contradictory. This practice delves into the idea of reconciling diverse elements, thoughts, or realities to coexist in synergy or synchrony (Nicolescu, 2002). At its core, compossibilizing is about considering the possibles that emerged in the previous step of the method ("profusing") and weaving as many as possible of these into a coherent, integrated structure, suggesstive of a better possible world aspiring to be realized.

Instead of viewing possibilities as agonistic entities, this practice encourages us to see many of them as interwoven threads of a larger and more or less sustainable narrative. It's about finding unity in diversity, about reconciling the seemingly irreconcilable (Robinson, 2001). The genesis of compossibilizing can be traced back to ancient holistic philosophies that emphasized the interconnectedness of all things. From the Daoist belief in the natural balance of Yin and Yang (Slingerland, 2007) to the Indian philosophy of Advaita that speaks of non-dualism (Sarma, 2011), the essence of compossibilizing is deeply rooted in the human psyche and in nature's complex movement toward unity and symbiosis in multivalence.

Once again, we need here to practice a form of deep listening to the symphony of life, discerning the myriad notes, and orchestrating them within our practices and projects (Oliveros, 2005). Compossibilizing also urges us to embrace apparent contradictions, to find beauty in creative dialectic paradoxes. In a world that often demands black and white answers, this practice reminds us of the myriad shades of combined colors that add depth and richness to our experiences. Ultimately, this is more than a practice; it's a worldview and it can become a transformative mindset (Dweck, 2006). It beckons us to nurture a multivalent way of thinking and imagining that celebrates the collaborative potential of the multiverse as a synergic whole, in which many possible worlds might be intertwined (Greene, 2011). Through compossibilizing, we train ourselves to not just navigate the world as spectators or consumers; we become active co-creators of the best of all compossible worlds, ethically aware that, even if anything seems possible in theory or locally, in practice and in the long-term not everything is compossible if we are to be vigilant about the eudynamic balance between persons, the earth, our ways of working and living in societies.

The seventeenth-century mathematician and thinker G.W. Leibniz, whose name appears throughout this book, introduced the concept of compossibility to address a profound philosophical question: given the infinity of possible worlds that God (or the Creal) might have created or is continuously creating, what principle guided the cosmic selection of our particular world? Leibniz's answer was compossibility, the idea that a world may represent "the best of all possible worlds" not because it is ideal, but because it optimizes the harmony of its diversity (Leibniz, 1989). To make a given world, you need possibles that are never fully contradictory. For Leibniz, a world is a system that thrives to optimize the greatest possible diversity while maximizing existence, which is the actualization of potentials. For centuries we have been champions at possibility thinking, generating what seemed like local miracles. But we neglected to think about the impact of our inventions upon the world as a holistic system, which is the main task of thinking compossibly.

Compossibilizing seeks not static harmony but what we might call "eudynamia"—the healthy dynamic compatibility of flowing or explosive forces. Like fireworks in choreographed sequence, different possibilities can detonate together in patterns that enhance rather than cancel each other's creative force. Between flow and blast lies what we might term "pulse"— rhythmic phenomena that create patterns without losing their essential creative energy. Compossible worlds pulse with coordinated creativity, each element contributing to an overall and ongoing symphony. In the symphony of live, there is no coda.

Leibniz suggested that reality unfolds through continuous monadic metamorphoses rather than abrupt breaks, a view remarkably consonant with our contemporary understanding of quantum probability waves and evolutionary development. Leibniz further defined existence as "compossible with the most perfect," where "more perfect" means "that which has more of reality or of positive entity" (Sleigh, 1983). Compossibility, according to Leibniz, is not merely the absence of contradiction but the actual presence of compatibility.

While contemporary thought has largely moved beyond Leibniz's theological framework, his insight about compossibility remains profoundly relevant in an age facing complex systemic challenges. In an interconnected world where local actions generate global consequences, the ability to create compossible systems—compatible within themselves and with their larger contexts—becomes essential for sustainable innovation. The philosopher Gilles Deleuze revived interest in Leibnizian compossibility, seeing it as a way to understand the "folding" of reality into complex patterns without reducing its intrinsic multiplicity (Deleuze, 1992). For Deleuze, compossibility doesn't eliminate difference but organizes it into more or less functional harmonies.

Leibniz had the idea that compossibility is a state of non-contradiction between two matters of fact. But how do we evaluate compossibility between two disparate things, say a violin and a bathtub? Of course, we can examine the laws of matter that keep the atoms together, probably comparable in the case of the bathtub and the violin. But these are two objects that have a specific use, one to play music, the other to take a bath. These two actions do not seem incompatible, and they could even be performed simultaneously (though possibly with some risk to the instrument). Beauty is sometimes a strange and disturbing matter in our world, as French poet Lautréamont—who lost his mind thinking about the disparity of our experiences—wrote in *The Songs of Maldoror*:

> *He is handsome like the retractility of the claws in birds of prey; or, yet again, as the uncertainty of muscular movement in sores in the soft spot of the posterior cervical region; or, rather, as the perpetual motion rat-trap which is always reset by the trapped animal and which can go on catching rodents indefinitely and works even when it is hidden under straw; and, above all, as the fortuitous encounter of a sewing machine and an umbrella on a dissecting table!*

Our worlds are so diverse that one may be overwhelmed, like Lautréamont, by the cacophonic horror of it all, and wonder how to generate compossibility. This is not merely about logical non-contradiction but about the quality of the relationships between elements. Some relationships are complementary, for instance when opposites complete each other. Others are supplementary, where elements add to each other. Still others are synergistic, where the interaction produces something greater than the sum of parts. The art of compossibilizing involves discerning which kinds of relationships might exist or be pruned among the diverse elements generated during profusing.

Remember that physicist David Bohm (1980) proposed the concept of "implicate order" to describe how apparently contradictory phenomena at the explicate (visible) level may be harmoniously integrated at a deeper implicate level. Similarly, in compossibilizing, we look beyond surface incompatibilities to discover deeper patterns of potential dynamic harmony. This doesn't mean forcing artificial connections but rather discovering where genuine complementarity might exist or evolve. Developmental psychologist Robert Kegan's (1994) research on orders of consciousness shows how we naturally evolve toward more integrative forms of meaning-making across the lifespan, with mature consciousness characterized by the ability to hold and integrate multiple perspectives without reducing them to a single frame. Perhaps this process can be democratized and accelerated via the crealectic method, such that we don't gain wisdom when it is too late to act.

When we compossibilize, we are not always eliminating contradiction but also transforming it into creative tension. Architect Christopher Alexander et al. (1977) described how living spaces achieve harmony and a sense of living wholebess not from eliminating all tension but from creating coherent patterns and by resolving tensions at one scale while maintaining productive tensions at other scales. This evolving gradient of harmony creates spaces that feel alive precisely because they balance resolution at some levels and with dynamic relationships at others. Similarly, in compossibilizing we seek not a flat homogeneity but a rich, multilayered dynamic harmony that preserves diversity while creating functional coherence. This is eudynamia, a good relationship with the possible and the compossible. Alexander's monumental investigation into the nature of life in built environments identifies fifteen fundamental properties that create what he calls "living structure," patterns that enhance life rather than diminish it (Alexander, 2002–2005). These properties, including "levels of scale," "strong centers," and "good shape," provide concrete principles for compossibilizing that ensure emerging wholes possess what Alexander calls "the quality without a name"—a difficult-to-define but unmistakable aliveness.

These questions lead us to consider what philosophers call ontological pluralism, the idea that there might be multiple valid ways for things to exist (James, 1996). Some aspects of our compossible worlds might be characterized by stability and robustness (like physical laws or ethical principles), while others might be deliberately fluid and evolving (like cultural expressions or technological applications). Anthropologist Arturo Escobar (2018) describes how indigenous cosmologies often embrace this kind of pluriversal thinking, where different modes of existence interrelate without requiring uniform categories or principles. The crealectic approach similarly embraces a pluralistic compossibility that doesn't force artificial unity but discovers resonances and complementarities across different domains and modes of existence. And as per the CIPHER model and the principle of integrated 4E cognition (see introduction), the locus of compossible dynamics is at the intersection of our personal embodiments, our extended belonging to the earth, our enactive ways of working, and our embedded participatory sense-making in societies.

Agents and Structures of Compossibility

The question of who decides what to compossibilize raises important issues of authority, expertise, and collective wisdom. In many conventional approaches to innovation, decisions about compatibility are made implicitly by those with the most power or the loudest voices. The crealectic method attempts to democratize this process while still recognizing the value of expertise in different domains.

When we apply the crealectic method in organizational contexts, we structure the compossibilizing phase to balance individual insight with collective intelligence. This might involve techniques like "affinity clustering," where participants individually identify connections between ideas and then collaboratively refine these clusters, or "worlds creation," where small teams each develop a compossible model and then share insights across teams. These approaches prevent domination by single perspectives while still allowing expertise to inform the process appropriately.

The question of expertise raises an interesting paradox: those with deep expertise in a domain may have difficulty imagining radical innovations precisely because their knowledge has been shaped by existing paradigms, and they may be locally rewarded to perpetuate old ways of doing. Innovation historian Andrew Hargadon (2003) shows how breakthrough innovations often come from individuals or teams that bridge multiple domains rather than from single-domain experts. The compossibilizing phase deliberately creates space for this kind of cross-domain integration. This is a reflection for us on how things connect, and how things tolerate themselves. Cultural theorist Homi Bhabha (1994) describes how creative cultural innovations emerge in what he calls "third spaces," hybrid zones where different cultures meet and negotiate meanings and identities beyond simple domination or assimilation. The compossibilizing phase creates such third spaces, where different possibilities can interact without predetermined hierarchy.

Compossibilizing is less about "what if" than about "what with"? How do things cohabit? Feminist philosopher Donna Haraway (2016) provocatively proposes "staying with the trouble" and the inherent ambivalence of human beings rather than seeking neat resolutions or transcendent solutions. Compossibilizing doesn't avoid difficult tensions but rather explores how these tensions might be holistically productive rather than destructive. Systems scientist Donella Meadows (1999) identified leverage points in complex systems, high potential places where small shifts can produce large changes. In compossibilizing, we look for such leverage points where relationships between possibilities might be transformed, from contradiction to complementarity, from conflict to synergy. This requires what psychologist Howard Gardner (2006) calls "synthesizing minds," the ability to integrate diverse elements into coherent wholes without losing the distinctive value of each component. Compossibility thinking cultivates this synthesizing capacity through deliberate practice in finding connections and complementarities across domains.

A peaceful army is moving slowly, like ants forming a column of workers to edify the cathedral that they will inhabit. Some patterns favor viability and long-term possibility. What are the fundamental elements that are needed for two possibles to coexist, that is to express the maximal degree of their potential into the external world? Worlds are not simple objects, they are our co-creation, and we can participate in their constitution if we are not prevented to do so. Compossibility is about a living structure, we are looking for an archi-structure, a holistic pattern for living structures that maximizes their domain of possibility in a way that does not injure the whole. The com- of compossibility does not mean only together but also over-structure, a healthy compatibility of constitution.

What is a structure? It is an invitation to acquire a shape, therefore it is a filter of multiplicity. Patterns are not just repetitions; they are molds, or embodied algorithms. Things get tried once, twice and several times, and that repetition starts to create a spacetime, a territory, a virtual shape. Think about growing organisms: they follow a pattern from embryo to full deployment, they unfold according to a program that is flexible to variations and yet robust. There is a necessity in most developments that is not of mere willpower, and such necessity is also what we are looking for in the compossible mindset. It's a dialectic of creativity and necessity.

Mathematician Roger Penrose has explored how complex structures emerge from simple patterns through processes he calls "quasiperiodic tilings" (Penrose, 1989). These structures exhibit order without simple repetition, much like how compossible worlds might integrate diverse elements through higher-order patterns. Biologist Stuart Kauffman's research on self-organization in complex adaptive systems shows how order can emerge spontaneously in systems with the right degree of connectivity (Kauffman, 1995). This suggests that compossibilizing isn't only about imposing order externally but rather discovering and enhancing the natural tendency of diverse systems toward self-organization, when its possibles are properly connected.

Psychologist Carl Jung proposed the concept of the collective unconscious, a shared structure of archetypes and symbols that may shape individual experience while transcending it (Jung, 1959). Similarly, in compossibilizing, we might discover structural values that transcend individual ideas while giving them coherent relationships. These patterns aren't imposed but both generated and discovered through attentive engagement with the material offered in profusing. Here again we are talking about what Alexander identified as "pattern languages," recurring design solutions that solve similar problems across different contexts (Alexander et al., 1977). In compossibilizing, we seek such pattern languages that might connect diverse possibilities in dynamic and functional harmonies.

Systems theorist Fritjof Capra (1996) distinguishes between pattern (the configuration of relationships), structure (the physical embodiment of pattern), and process (the activity that creates and sustains pattern). Compossibilizing engages all three levels: identifying patterns of potential relationship, exploring potential structures that might embody these patterns, and anticipating processes that could actualize these structures. This multilevel approach may ensure that compossible worlds aren't merely conceptual but have pathways toward realization and implementation (see Chap. 6).

This raises an interesting tension between the potential and the actual. Bergson (2007) distinguished between "closed" and "open" systems, the former being deterministic and the latter creative. Similarly, compossibilizing seeks to create "open structures," frameworks that provide coherence without rigid determination, allowing for ongoing creativity within a constantly re-equilibrated system. This aligns with what complexity theorist Edgar Morin calls "organized complexity," where order and disorder work together rather than against each other (Morin, 2008).

Whitehead proposed that reality consists fundamentally of "actual occasions," momentary experiences that weave together past influences into new unities (Whitehead, 1978). The world is not made of static substances but of creative events that continuously integrate diverse influences. Compossibilizing mirrors this process in its inspiration, articulating together diverse possibilities into new unities without reducing their distinctive qualities.

Practical Approaches to Compossibilizing

Can humanity think beyond its own symbols and familiarity of concepts? Can we innovate in leaps, or can this only be done incrementally, one step at a time? There are on the one hand specialists, for instance of an aspect of energy systems, and on the other side there are managers or leaders who might not be detailed experts on the field. How do we connect their insights? This can be done via the workshops themselves, which means crealectics is a place for cross-pollination, displaying a spacetime for interdisciplinary practitioners to learn from each other.

Invoking again our Vattenfall R&D study case, let's suppose we are holding a workshop about the constitution of a healthy and holistic energy system, and that during the profusing stage, the following ideas—among others—are produced, such as:

> *Community solar microgrids, meditation-enhanced conductivity, waste heat recovery systems, dance floor power generation, local food energy networks, fully-electrified public transport, grief-to-energy conversion sanctuaries, dream harvesting sleep pods, geothermal district heating, energy literate communities, tidal power installations, regenerative agriculture practices, collective joy battery farms, biorhythm-responsive power grids, breath-powered personal devices, mycelium energy transfer systems, wildlife collaboration corridors, interspecies energy communication, etc.*

A first way to compossibilize them could be to organize these ideas into structurally compatible categories that show how they could work together in a holistic energy system:

- Infrastructure & Generation Systems: These components form the physical backbone for producing, storing, and distributing energy.
- Energy Efficiency & Material Flows: This group may focus on optimizing energy usage and creating circular material pathways to minimize waste.
- Community & Governance Structures: These elements address social organization and decision-making frameworks essential for energy system transformation.
- Biological & Natural Integration: These approaches incorporate and curate living systems and natural processes into energy networks.
- Alternative Energy Capture: This category represents innovative approaches to generating energy from human activities and states.
- Responsive & Adaptive Systems: These components enable the energy system to respond dynamically to natural or artificial rhythms.

From here, we can create for instance two clusters of compossible ideas within this list, making groupings of elements to demonstrate how compossibilizing creates coherent worlds:

Cluster 1: Technologically Centered Resilient Infrastructure. This cluster would integrate technical solutions with community governance to create resilient local energy systems. Core Infrastructure Elements such as

Community solar microgrids or Smart home energy monitors; supporting protocols or systems such as Energy literate communities or Biorhythm-responsive power grids. This cluster anticipates a technically robust, community-governed energy ecosystem centered on distributed renewable generation with smart management systems. The integration of energy literacy, governance structures, and responsive grids may ensure optimization and democratization of energy resources.

Cluster 2: Biologically Integrated Living Systems. This cluster would merge biological processes with alternative energy approaches to create a nature-aligned energy paradigm. Core biological elements such as Mycelium energy transfer systems or Regenerative agriculture practices; supporting components such as Local food energy networks or Dance floor power generation. This cluster may create a living energy system that harmonizes with natural cycles and human biological rhythms. By integrating food production, natural materials, and conscious energy generation, it establishes a regenerative approach that enhances rather than depletes ecological systems.

The fundamental structure of cluster 1 is a decentralized technical network governed by community-based decision systems. At its core, this model integrates distributed energy generation and storage with intelligent management systems. The essential components create a feedback loop where physical infrastructure provides resilient energy production and storage; smart systems optimize resource allocation and consumption pattern; community governance ensures democratic control and transparency; education and standards maintain system integrity and improvement. This structure establishes a self-regulating technical ecosystem that balances efficiency with community autonomy.

The essential structure of cluster 2 is a regenerative ecological web that merges biological processes with human consciousness. This model functions as a living organism where biological systems provide energy generation and transmission pathways; natural cycles determine storage and usage patterns; human-nature interfaces enable reciprocal energy exchange; conscious participation and sense-making activates energy flows beyond mechanical processes. This structure creates a symbiotic relationship between human settlements, energy systems, and the broader ecological context, functioning less as infrastructure and more as a living extension of natural systems.

These two clusters are compossibilized examples that represent different but potentially complementary approaches to holistic energy systems.

One emphasizes technical resilience through distributed systems and community governance, while the other prioritizes biological integration and consciousness-based participation. A fully compossible approach might find ways to integrate elements of both, creating what systems theorist Joanna Macy (2007) calls "mutual causality," where technical and living systems enhance rather than obstruct each other. The intelligent management systems of Cluster 1 could learn to respond to the biological rhythms and regenerative cycles of Cluster 2, while the living energy networks of Cluster 2 could benefit from the democratic governance structures and resilient infrastructure of Cluster 1, generating hybrid possibilities where smart grids become extensions of mycelial networks and community decision-making processes mirror natural ecosystem feedback loops.

As the above exercise demonstrates, the compossibilizing phase doesn't simply reduce diversity but can organize it into coherent patterns that preserve the richness of different approaches while creating functional relationships between them. These patterns aren't arbitrary but emerge from careful attention to the inherent properties and potential relationships of the ideas generated during profusing. Of course, due to lack of space, the above example remains speculative and somewhat abstract. It would benefit from reiterations of refinement and clarification in order to become at least partially implementable. But it is the role of the next phase in the crealectic method: realising, to actualize and concretize the compossible hypotheses (see Chap. 6).

From Compossibilizing to Realizing: The Bridge to Action

The multiverse is an invitation to healthy freedom; one can become a universe jumper, leaving possibles in a state of connected alliance, a vibration of connections, a symbiosis between several of many worlds, such that bridges are created between cosmoses to avoid the rigid production of the same. Crealectics is a practice that honors the multiversal nature of creation, allowing seemingly disparate ideas to be held in beneficial tension. This natural approach of intercreation resonates with what physicist and philosopher Karen Barad (2007) calls "intra-action," where entities don't pre-exist their relationships but emerge through them. In compossibilizing, we're not simply connecting pre-existing ideas but allowing new possibilities to emerge through their relationships.

Philosopher William James (1996) proposed the concept of the "pluriverse" to challenge the assumption that reality must ultimately reduce to a single unified system. Similarly, compossibilizing doesn't seek to reduce all possibilities to a single world but rather to discover productive relationships between different possible lifeworlds. Anthropologist Eduardo Viveiros de Castro's (2014) work on "multinaturalism" suggests that indigenous cosmologies often embrace multiple natures rather than multiple perspectives on a single nature. This radical multiplicity reflects what philosophers call "contingency," the awareness that things could be otherwise and that alternatives always exist (Solnit, 2004).

At the same time, compossibilizing isn't always relativistic nor does it claim that compossibility is arbitrary. It seeks what nineteenth century philosopher Charles Sanders Peirce called "concrete reasonableness," patterns of relationship that work in practice while remaining open to revision and development (Peirce, 1998). This pragmatic approach aligns with what sociologist Bruno Latour calls "compositionism," the careful composition of shared worlds through negotiation rather than through appeal to universal principles or pure relativism (Latour, 2010).

We are plunging our minds into the very idea of connection and interconnection, not the things in themselves but their interconnection. Look at how the very disparate on earth is made to cohabit! Lifeworlds are experienced universes. Every form of life is a lifeworld. Compossible worlds are a composition of lifeworlds that are symbiotic or eudynamic. Compossibilizing is not merely a technical process but an ethical and political one. Whose worlds are included or excluded? Whose possibilities are deemed compatible or incompatible? What is evolving faster or slower? These decisions have profound implications for justice, sustainability, and planetary flourishing. Environmental philosopher Val Plumwood (1993) critiqued what she called "hyperseparation," the tendency to create artificial divisions between humanity and nature, mind and body, reason and emotion. Compossibilizing challenges these separations, seeking integration without reducing difference to sameness. Everything that is embodied, embedded, extended and enactive is a living agent in the eudynamic system we call our world.

Compossibilizing requires careful attention to the relationships between possibilities, not just to their individual properties. This care extends to what science studies scholar Isabelle Stengers (2015) calls "the cosmopolitical," where diverse beings—human and non-human—may negotiate shared worlds without imposing uniform standards or values. There must be a benefit in slow growth, slow innovation. The erratic and ambivalent living rhythm is a laborious conjunction of inspiration and stupor, trial and error, moments of enthusiasm and moments of nothingness. Duration is elaboration.

This temporal dimension of compossibilizing connects with what philosopher Jacques Derrida called "différance," the continuous deferral and differing that constitutes meaning (Derrida, 1982). Compossibilizing doesn't seek immediate closure but remains open to ongoing development and revision. Moreover, crealectic sessions must try and create space for constructive divergence of views. This is easier said than done. It would require, as we mentioned earlier, what Fricker calls "epistemic justice," recognizing and listening to the knowledge and insight that comes from different positions and experiences (Fricker, 2007).

As diverse possibilities coalesce into patterns of potential compossibility, we naturally begin to sense pathways toward realization. The compossibilizing phase doesn't conclude with abstract patterns but with emerging structures that have the potential to manifest in concrete reality. This transition from compossibility to realization resembles what design theorist Donald Schön called "seeing-moving-seeing," a recursive process where perception leads to action, which in turn transforms perception (Schön, 1983).

The compossibilizing phase does not need to end with a single "correct" pattern but can point to several potential worlds, each with its own internal coherence and relationships to other possible worlds. These miniworlds aren't theoretical abstractions but potential pathways for concrete action. The compossibilizing phase concludes not with final answers but with drafts or conceptual simulations of what innovation researcher Michael Schrage calls "prototypes," provisional manifestations that invite further development and refinement (Schrage, 2000).

In workshops, this transition might involve creating visual maps or narrative scenarios that illustrate compossible futures, identifying initial projects that could begin bringing these futures into being, and exploring what philosopher Donna Haraway calls "response-ability," the capacity to respond to emerging possibilities rather than simply implementing predetermined plans (Haraway, 2016). This preparedness for response also aligns with what complexity theorist Dave Snowden calls "safe-to-fail experiments," small initiatives designed to test and develop promising possibilities without requiring large-scale commitments before learning occurs (Snowden & Boone, 2007).

As we transition from compossibilizing to realizing, we carry forward not just specific ideas or patterns but what anthropologist Tim Ingold, inspired by Deleuze, calls "lines of becoming," trajectories of development that remain open to evolution rather than predetermined blueprints (Ingold, 2015). These lines of becoming connect the compossibilizing phase to the realizing phase, ensuring that implementation doesn't abandon too many of the rich possibilities generated through the creative process but rather tries to manifest them in forms that preserve their creative potential.

Conclusion: The Harmony of Multiplicity

Compossibilizing is a verb, an act, something that we do actively. In fact we compossibilize every day in our daily lives, trying to adjust and mend the disparate, the diverse, the unimpossible. The brain is always trying to create meaning and compossibilizes in the night while dreaming. Perhaps dreaming is our mental training to be able to compossibilize: if we can make a story out of the many different impressions and affects and percepts of a dream, then, with some cooperation, we should be able to connect in real life the equally disparate data of our shared experience.

Deleuze called "transcendental empiricism" the exploration of the conditions that generate actual experience rather than merely analyzing experience after it occurs (Deleuze, 1994). Compossibilizing operates at this transcendental empirical level, exploring not just what is but what could be connected, and how different possibilities might cohere in new patterns of eudynamic realities. Local improvements are less realistic than is usually claimed, because they are often blind to their global consequences. Planetary sustainability demands a universal and therefore a philosophically pragmatic stance. This aligns with what business theorist C.K. Prahalad called "next practice" rather than "best practice," innovation that creates new and essential patterns rather than merely optimizing existing ones (Prahalad & Hamel, 1990).

The compossibilizing phase stands as the crucial moment of the crealectic method, where the abundance of possibility generated in profusing begins to find coherent patterns without being reduced to uniformity. It embodies what philosopher Theodor Adorno called "negative dialectics," a form of thinking that preserves difference within unity rather than subsuming particulars under dogmatic concepts (Adorno, 1973). This preservation of multiplicity within harmonization distinguishes compossibilizing from approaches that seek unity through reduction or elimination of difference.

Is compossibilizing love? It is trying to unite the disparate into one world, one multiverse, a multivalent universe. Philosopher Hans-Georg Gadamer called "fusion of horizons" the merging of different perspectives into more comprehensive understanding without eliminating their differences (Gadamer, 1960/2004). Compossibilizing creates such fusions, not by reducing all perspectives to a common denominator but by discovering how diverse perspectives might illuminate different aspects of a complex whole. Aldo Leopold called it "thinking like a mountain," adopting a holistic perspective that integrates multiple timeframes and spatial scales into a more complete understanding (Leopold, 1949).

Whitehead suggested that the ultimate irony of the universe is the fact that it creates particulars through its universal creativity, and universals through its particular manifestations (Whitehead, 1978). Compossibilizing participates in this cosmic irony, seeking patterns that unify without eliminating the particularity that makes each possibility unique. This ironic awareness prevents compossibilizing from becoming dogmatic or reductive, maintaining openness to ongoing creativity within evolving structures.

Many questions remain that cannot be addressed in a small methodological handbook. What is our model when we compossibilise and is the person-earth-work-society eudynamic quadrant enough? What are the logical nodes that we use when we connect the dots? Can we ever measure eudynamia and should we try to? These questions remind us that compossibilizing is not a mechanical process but a creative art in itself. It requires what philosopher Martha Nussbaum calls "poetic justice," the ability to empathically imagine how diverse elements might relate without imposing predetermined categories or values (Nussbaum, 1995). The compossibilization phase should be thought as including the more or less poetic observer in the equation.

This exercise of thinking compossibly rather than just possibly should be often reproduced, so as to create a culture or second nature. To *think compossibly* means to open our eyes and think about relationships rather than about just actualization and realization of a local technological or productive possibility. What is left out of the possible may be as important as what is assembled in.

These reflections on methodology connect with what French philosopher Michel Foucault called "technologies of the self," practices through which individuals transform themselves in relation to social norms and possibilities (Foucault, 1988). Innovation and innernovation or self-innovation should be cultivated together. In practice, I almost always combine the crealectic method with the SMILE_PH method (Sense-Making Interviews Looking at Elements of Philosophical Health), to explore in individual sessions the person's sense of the possible, but also other senses such as sense of self and sense of purpose (de Miranda, 2024). Beyond corporations or institutions, compossibilizing can function as such a technology of the self, helping a person develop new capacities for integrative thinking and relationship-building across biographical differences. Eudynamia is an existential ideal, not only an organizational one. This developmental dimension echoes what education theorist Jack Mezirow called "transformative learning," changes in perspective that alter not just what we know but how we know (Mezirow, 1991).

As we conclude this exploration of the compossibilizing phase of the crealectic method, we recognize that it stands as both a practical exercize and a philosophical stance toward reality. It embodies what philosopher Isabelle Stengers calls an "ecology of practices," a way of relating diverse approaches without reducing them to a single standard or hierarchy (Stengers, 2005). In an age increasingly defined by fragmentation and polarization, compossibilizing offers a path toward the capacity to engage productively across differences without either eliminating them or allowing them to degenerate into destructive conflict (Mouffe, 2000). Leibniz was often mocked, for instance by Voltaire, for speaking of the best of possible worlds. Now that our world is becoming global, his concept of compossibility is the tool we need to co-create the first planetary civilization.

The compossibilizing phase of the crealectic method has prepared us for the final phase: realizing. Having generated abundance through profusing and discovered patterns of potential eydynamic harmony through compossibilizing, we are now ready to bring these compossibilities into concrete manifestation and implementation, without which there is no real innovation. As we move into this final phase, we carry forward not just specific ideas or structures but a transformed relationship to possibility itself—a capacity to see the world not as a fixed reality to which we must adapt but as a fluid field of potential in which we actively co-participate. This participatory consciousness, cultivated through the entire crealectic journey, becomes the foundation for the realizing phase, where possibility and actuality meet.

References

Adorno, T. W. (1973). *Negative dialectics* (E. B. Ashton, Trans.). Continuum. (Original work published 1966).

Alexander, C. (2002–2005). *The nature of order* (4 vols.). Center for Environmental Structure.

Alexander, C., Ishikawa, S., & Silverstein, M. (1977). *A pattern language: Towns, buildings, construction.* Oxford University Press.

Barad, K. (2007). *Meeting the universe halfway: Quantum physics and the entanglement of matter and meaning.* Duke University Press.

Bergson, H. (2007). *Creative evolution* (A. Mitchell, Trans.). Palgrave Macmillan. (Original work published 1907).

Bhabha, H. K. (1994). *The location of culture.* Routledge.

Bohm, D. (1980). *Wholeness and the implicate order.* Routledge.

Capra, F. (1996). *The web of life: A new scientific understanding of living systems.* Anchor Books.

Castro, E. V. (2014). *Cannibal metaphysics: For a post-structural anthropology.* Univocal Publishing.

De Miranda, L. (2024). *Philosophical Health: a Practical Introduction.* Bloomsbury.

Deleuze, G. (1992). *The fold: Leibniz and the Baroque.* University of Minnesota Press.

Deleuze, G. (1994). *Difference and repetition* (P. Patton, Trans.). Columbia University Press.

Derrida, J. (1982). *Margins of philosophy* (A. Bass, Trans.). University of Chicago Press.

Dweck, C. (2006). *Mindset: The new psychology of success.* Random House.

Escobar, A. (2018). *Designs for the pluriverse: Radical interdependence, autonomy, and the making of worlds.* Duke University Press.

Foucault, M. (1988). Technologies of the self. In L. H. Martin, H. Gutman, & P. H. Hutton (Eds.), *Technologies of the self: A seminar with Michel Foucault* (pp. 16–49). University of Massachusetts Press.

Fricker, M. (2007). *Epistemic injustice: Power and the ethics of knowing.* Oxford University Press.

Gadamer, H-G. (2004). *Truth and method* (J. Weinsheimer & D. G. Marshall, Trans.). Continuum. (Original work published 1960).

Gardner, H. (2006). *Five minds for the future.* Harvard Business School Press.

Greene, B. (2011). *The hidden reality: Parallel universes and the deep laws of the cosmos.* Knopf.

Haraway, D. (2016). *Staying with the trouble: Making kin in the Chthulucene.* Duke University Press.

Hargadon, A. (2003). *How breakthroughs happen: The surprising truth about how companies innovate.* Harvard Business School Press.

Ingold, T. (2015). *The life of lines.* Routledge.
James, W. (1996). *A pluralistic universe.* University of Nebraska Press. (Original work published 1909).
Jung, C. G. (1959). *The archetypes and the collective unconscious.* Princeton University Press.
Kauffman, S. (1995). *At home in the universe: The search for the laws of self-organization and complexity.* Oxford University Press.
Kegan, R. (1994). *In over our heads: The mental demands of modern life.* Harvard University Press.
Latour, B. (2010). An attempt at a "compositionist manifesto". *New Literary History, 41*(3), 471–490.
Leibniz, G. W. (1989). Philosophical essays (R. Ariew & D. Garber, Trans.). Hackett Publishing.
Leopold, A. (1949). *A Sand County almanac.* Oxford University Press.
Macy, J. (2007). *World as lover, world as self: Courage for global justice and ecological renewal.* Parallax Press.
Meadows, D. (1999). *Leverage points: Places to intervene in a system.* The Sustainability Institute.
Mezirow, J. (1991). *Transformative dimensions of adult learning.* Jossey-Bass.
Morin, E. (2008). *On complexity.* Hampton Press.
Mouffe, C. (2000). *The democratic paradox.* Verso.
Nicolescu, B. (2002). *Manifesto of transdisciplinarity.* SUNY Press.
Nussbaum, M. (1995). *Poetic justice: The literary imagination and public life.* Beacon Press.
Oliveros, P. (2005). *Deep listening: A composer's sound practice.* iUniverse.
Peirce, C. S. (1998). *The essential Peirce: Selected philosophical writings.* Indiana University Press.
Penrose, R. (1989). *The emperor's new mind: Concerning computers, minds, and the laws of physics.* Oxford University Press.
Plumwood, V. (1993). *Feminism and the mastery of nature.* Routledge.
Prahalad, C. K., & Hamel, G. (1990). The core competence of the corporation. *Harvard Business Review, 68*(3), 79–91.
Robinson, J. (2001). Squaring the circle? Some thoughts on the idea of sustainable development. *Ecological Economics, 48*, 369–384.
Sarma, D. (2011). *Classical Indian philosophy: A reader.* Columbia University Press.
Schön, D. A. (1983). *The reflective practitioner: How professionals think in action.* Basic Books.
Schrage, M. (2000). *Serious play: How the world's best companies simulate to innovate.* Harvard Business School Press.
Sleigh, R. C. (1983). Leibniz on the two great principles of all our reasonings. In M. Hooker (Ed.), *Leibniz: Critical and interpretive essays* (pp. 35–49). University of Minnesota Press.

Slingerland, E. (2007). *Effortless action: Wu-wei as conceptual metaphor and spiritual ideal in early China*. Oxford University Press.
Snowden, D. J., & Boone, M. E. (2007). A leader's framework for decision making. *Harvard Business Review, 85*(11), 68–76.
Solnit, R. (2004). *Hope in the dark: Untold histories, wild possibilities*. Nation Books.
Stengers, I. (2005). The cosmopolitical proposal. In B. Latour & P. Weibel (Eds.), *Making things public: Atmospheres of democracy* (pp. 994–1003). MIT Press.
Stengers, I. (2015). *In catastrophic times: Resisting the coming barbarism*. Open Humanities Press.
Whitehead, A. N. (1978). *Process and reality* (Corrected ed.). Free Press. (Original work published 1929).

CHAPTER 6

Realizing

Abstract Realizing encompasses both understanding and manifestation or actualization—the dual process of cognitive insight and practical implementation. This chapter develops the concept of "authoricity" (the integration of authenticity, authorship, and authority) and explores four dimensions of realization: cognitive, technical, social, and inner. It distinguishes between "Real Evidence" and "Creal Evidence," offering frameworks for holistic evaluation that honor both quantitative and qualitative aspects of creative achievement.

Keywords Realizing • Authoricity • Manifestation • Implementation • Real Evidence • Creal Evidence • Holistic evaluation • Practical wisdom • Ethical innovation • Transformative learning

L. de Miranda, *Crealectics as a Creative Method*,
Palgrave Advances in Possibility Studies,
https://doi.org/10.1007/978-3-032-00427-7_6

Fig. 6.1 The realizing phase symbol

We have cleaned our minds, traversed the ocean of possibility and became one with it, filtered and connected our impressions and intentions through the lens of compossibility, and now stand at the threshold of concrete manifestation. The creative journey culminates in this final phase of the crealectic method: realizing. It is about real world-making. Here we may transform the rich potentialities we have generated into tangible realities that can reshape our worlds. As we shall see, this phase embodies a mind-matter dual aspect that mirrors the very nature of creation itself (Fig. 6.1).

The Dual Nature of Realization: Understanding and Manifestation

The word "realizing" carries with it a fecund duality. On one hand, it signifies the act of coming to understand or perceive something clearly, like a revelation that involves both emotion and consciousness (Damasio, 1999). On the other, it embodies the action of making something real, of bringing an idea or vision to fruition and implementing it as a reflective practitioner (Schön, 1983). This dual essence offers a rich opportunity for insight into both our cognitive processes and our capacity for compossible action.

The journey of human evolution has been marked by moments of realization. From the intellectual awakening that led early humans to harness fire to the eureka moments of scientists and thinkers, realizing in the sense of understanding has been a beacon, guiding humanity through the corridors of progress (Diamond, 1997). Our individual biographies are similarly constructed of these grateful and sometimes simultaneously painful moments where we suddenly become aware of a blind spot, a pattern in our behavior, or a new existential possibility that transforms us.

But understanding alone, as potent as it might be, is here just one facet of realizing. Once an idea takes root in the mind, the next critical step may be to manifest it, to translate the abstract into the pragmatic and tangible (Dewey, 2005 [1934]; Brown, 2009). This transformational process is where dreams are sculpted into reality, where visions find form and substance, where innovation begins to influence our forms of life. Realizing is an act that can integrate mind and action to build a new worldly experience. The challenging interplay between understanding and actualizing is beautifully intricate, especially when the philosophical health of the system and the healing-growth future of humanity are at stake (de Miranda, 2021b). To realize is to take active care of and shape the eudynamic coherence between meanings, values, and actions, at the crossroad of personhood, earthly belonging, transformative work and social embeddedness.

The philosophical health perspective emphasizes the importance of inner realization, of understanding one's true purposes and the fundamental essence of thought in action (de Miranda, 2024). This inner understanding, this pragmatic wisdom grounded in dialogue is the key to manifesting outer realities that are asymptotically harmonious and collectivelly meaningful. In modern psychology and neuroscience, the dual nature of realizing finds resonance in the idea of coherence as a bridge between thinking and doing (Thagard, 2002). In fact, cognitive processes that lead to understanding are closely linked with our ability to visualize, plan, and execute actions that make concepts real and actions more effectual (Gilbert & Wilson, 2007). To think is to prospect. The neural pathways activated when we grasp a new concept are intricately connected with those that drive our constructive abilities (Dietrich, 2004).

Realization embodies both the sudden blast of understanding that never comes without much preliminary exercise (the "eureka explosion") and the following or simultaneous sustained work of manifestation and experimentation. The moment of insight might feel instantaneous, but bringing that insight into reality requires the patience to channel explosive creativity into concrete forms. This is where the crealectic method becomes fit for transformational outcomes in our everyday practices and work, provided one trusts it incremental long-term effect instead of expecting immediate gratification. Realizing is a muscle that needs to be trained regularly if it is to be both effective and enlightened in the intercreation of eudynamic realities.

Deep innovation is a slow rather than anxious transformation of shared realities. Implementation is usually incorporated in the process. There is a courage required to realize, in both senses of the word. This is the phase that people and corporations often find most challenging to embrace, because they perceive that's where the game becomes serious. It's not what Boltanski and Chiapello (2005) called "The New Spirit of Capitalism"—experiential fun to keep the good soldiers in groupthink and perpetuate blind spots—it's meant to be genuine transformation.

What does it *really* mean, to "realize"? It suggests we can contribute to the production or understanding of the real. What is real? Firstly, that which is repeated to the point of becoming a thing or a shared experience in a world. Realizing implies a repetition, the decision that we will hold onto something in our minds or in the world. A new repetition must replace an old repetition, and in the semantic theater of life, this starts with a new interpretation. Crealecticians keep in mind that these social repetitions can always be questioned and recreated.

Every world is a set of possibles made to be more or less compossible, more or less eudynamic; realizing means contributing to actualizing one world, because reality is that which distinguishes worlds, otherwise connected by the multiverse of Creal. The common foundation of worldmaking is potential, virtuality, ultimate possibility. Realizing creates a decoherence of the virtual into the real. But the real never fully leaves the land of the possible, it remains part of the Creal. All real can be virtual or potential for another real, and that is the first thing one must realize: everything can be a possible for another reality, for another world.

There is honesty, a form of mental courage, in realizing as simple interpretation, because we need to see the world as we understand it is, not as someone else told us it is. Moving on, we leave the realm of speculation or fiction, and we need to integrate difference, noise, resistance and impurity, and be patient enough so that things happen differently. In today's world, this means an investment of resources and sometimes irreversible decisions, while the previous steps of the method are relatively inexpensive in comparison. True realization can be scary for individuals and groups, because of their existential cost and risk.Like capturing lightning in a bottle, we seek to realize innovations that retain their original creative blast while functioning effectively—fluidly—in the world. But sometimes, implementations may feel like a giant compromise, by which we may need to abandon all the idealism generated in the previous phases of crealectic thinking. This is where great achievers are tested: anyone can *have* a dream. Few can offer it to the world such that it transforms the world without sacrificing the dream.

Who has authorship of what is realized in an organization? Could it be that people do not feel motivated because their authorship is not recognized? Or perhaps they do not have the authority to execute? Could it be that they do not have the authenticity to fight for what they believe is right? This is where the concept of authoricity becomes crucial in the leadership path of becoming a realizer, not just a preacher or a reacher.

Authoricity: The Matrix of Creative Agency

We hinted at it in our introduction and come back to it now. Authoricity is, I have proposed in the context of my philosophical counseling work, the conjunction and dynamic harmony between authenticity, authorship, and authority. It represents the fullness of leadership or creative agency that enables effective realization (Fig. 6.2).

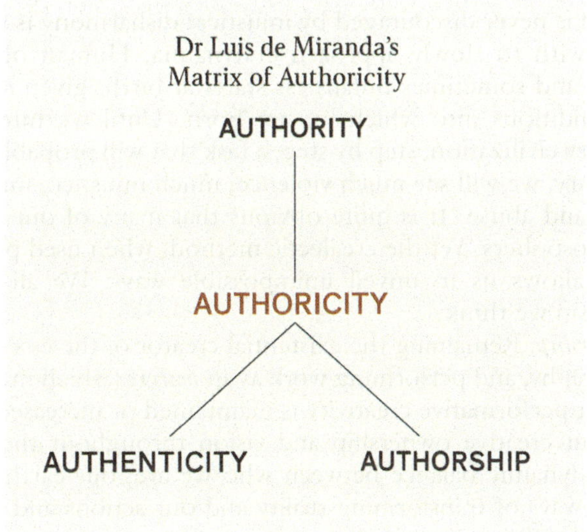

Fig. 6.2 The Authoricity Matrix

Three essential elements constitute authoricity:

1. Authenticity: Being genuine and faithful to what we believe is admirable and desirable. This involves honesty with oneself and others about values and intentions. Real authenticity is difficult to cultivate because we may confuse it with emotional reaction. This is where the SMILE_PH method helps (Sense-Making Interviews Looking at Elements of Philosophical Health), via sessions of personal dialogue along the following six elements: bodily sense, sense of self, sense of belonging, sense of the possible, sense of purpose and philosophical sense. To develop this form of philosophical counselling, I created the Philosophical Health International network. We can all be more authentic than we believe.

2. Authority: Having the courage and capacity to change things, to leverage obstacles, and to influence in terms of active power. This is the ability to mobilize resources and gain support for realization. Some people today are wary of the notion of authority because they associate it with coercion and abusive or unequal distribution of power. And it remains true in various contemporary institutions and corporations that those who have the highest power of decision may seem unfit for the complexity of the task, which is a polite designation for what the disempowered call them: *idiots* (which etymologically means that they do not really care about what we share but are mostly driven by ego trip). But a crealectician is never discouraged by injustice: disharmony is the material one works with to slowly approach eudynamia. Human obstacles will always exist and sometimes unfairness starts at birth, given the social or parental conditions into which we are born. Until we intercreate the first planetary civilization, step by step, a task that will probably take more than a century, we will see much violence, much injustice, much trauma, corruption and abuse. It is quite obvious that many of our "kings" are not yet philosophers. Yet the crealectic method, when used personally or in groups, allows us to unveil unimpossible ways. We all have more authority than we think.

3. Authorship: Remaining the existential creator of the essential aspects of our biography, and performing work as an *oeuvre* (which may be collective), so that performative creativity is maintained or increased over time. This is about creative ownership and vision throughout the process of realizing eudynamic balance between who we are, our earthly extended nature, our ways of transforming reality and our actions and reactions in society. We are all creatures and creators.

In order to be a great realizer, a person or a group must progressively exhibit a relatively high level of authoricity on all three factors. This might take decades of patient labor, and it's fine to feel that, here and now, one may be low in authoricity; it is an invitation to grow harmoniously. This triangle of creative agency provides a framework for understanding why some deep innovations are successfully realized while others remain mere possibilities. Without reflective authenticity, realization becomes hollow and more or less dishonest; without wise authority, it remains reactive and may act in bad faith; without authorship, it loses its distinctive vision and neglects the higher purpose. Investor Hemant Taneja's analysis of responsible innovation explains that the most admirable companies—perhaps *successful* if we redefine successful in contradistinction with mainstream capitalistic parameters—are what he calls "intended consequence" organizations, those that explicitly design for positive social and environmental impact from inception rather than treating such concerns as afterthoughts or communication assets (Taneja, 2022). This approach might embody the authoricity principle by integrating authenticity (genuine commitment to positive impact), authorship (creating responsibly for consequences), and authority (collectively distributing the agency to build systemic change). One might feel that the definitions of the three axes of the matrix could almost be substituted for one another. That is because there is a point in the matrix where authority, authenticity and authorship become one and the same quality: authoricity emerges.

Some of the crealectic workshops may lead to no decision or actual realization. The authoricity model helps explain why many ideas are not implemented or executed. Managers may be too low in one or more components of this triangle, or the chain of command may be too hierarchical and bureaucratic, resulting in unrealized potential, unclarity, and indecision. In some cases, that is fine, because not every idea needs to be realized, and sometimes inspiration or daydreaming are enough because they may lead to something else in the longer term. But it can be frustrating when promising innovative initiatives repeatedly fail to become real due to deficiencies in personal authoricity. The first thing to do here is to arrange by design for a repetition of workshops, such that change may emerge from reiteration. The second thing to do is to cultivate philosophical health in individuals with personal sessions of dialogue, thus combining the two methods of the CIPHER loop.

As mentioned earlier, the so-called concrete reality is the effect of repetitions, semantic, organic, routine-based, institutional, protocolar. Entwined realities may pertain to a cluster of lifeworlds, for instance the spacetimes experienced by agents. The problem of living in one world only is that we get used to everything being as it is and may naturalize habits into a fixed necessity rather than the effect of a repeated contingent institution. Everything becomes too rigid for hyper-realists, or monotonous, and they are only surprised by (and constantly on the lookout for) what dramatically changes and is particularly salient. In fact, we should be able to see movement, becoming and sparkling everywhere, the phosphorescence or blast of creation in every bit of life: every little thing is not just a *realia*, an object, but often a *crealia*, a potential.

The mind is participating in the enchantment; wonder is not out there only—we need to co-author the admiration. Some say that consciousness collapses the wave function, but it also reawakens the wave function, transforms the real into Creal, the inert into potential. Systems thinker Donella Meadows (2008) used to say that she saw feedback loops everywhere. To become a crealectician is about seeing *crealia* everywhere, bits of creation in progress, not just the *realia*, sets of things. To *realize* is to make a device that ensures that points of possibility remain offered to the future, and in doing so manifesting the grace of compossibility in an operational way.

Our entire economy is a fetishization of the real as a space of things. What is sold is a product, sometimes a standardized experience. But what is the most real? The spaces between things: the emptiness that allows us to desire, to imagine, to create.

The real as we know it or experience it is often a simplification based on utilitarianism. We are producing economically what people are made to believe they want. There is a risk that the process of realizing destroys part of the richness and potential of the previous steps of the crealectic method. To avoid that, we need a form of realizing that is not an impoverishment of our daily experience and of the availability of creative experience. At the same time, this should not lead to paralysis: authoritative decisions will occasionally be taken and the Gordian knot cut.

Inner-Novation: The Heart of Holistic Realization

Process philosophies insist that it is a fallacy to think that most things are solid and written in stone. We need to look at flows, blasts, relationships and attractions. Realizing means to produce holistic innovation,

innovation at the level of the whole, which means that emergence remains possible and is taken into account by design. Creation is an inner phenomenon, and therefore holistic innovation is also, as we have argued throughout this book, "inner-novation."

Such is the innovation we are bringing about as crealecticians: an encouragement for others to embrace their crealectic intelligence. Whatever we realize needs to favor that inner intercreative richness rather than impoverish it. Diversity must be the allowance to inner-create not just a multiverse, but a pluriverse of different dimensions: Person, Earth, Work and Society. Even in deterministic worlds, one can jump from one universe to another to create freedom and bifurcate paths. Eudynamia results from the dynamic interplay of the integrated dimensions of experience: embodied, extended, embedded and enactive. Eudynamic realities emerge through the continuous, recursive interactions between these four dimensions of experience, where embodied awareness (our lived, physical presence) dynamically shapes and is shaped by our extended cognition (how we think through tools, environments, and technologies), our embedded social contexts (the linguistic, cultural, relational, and institutional networks we inhabit), and our enactive engagement (the way we bring forth meaning through purposeful action in the world). This interplay is not merely additive but transformative—each dimension amplifies and modifies the others in an ongoing process of mutual specification, creating emergent possibilities that exceed what any single dimension could generate alone. The result is a form of flourishing that is simultaneously grounded in bodily wisdom, naturally or technologically augmented, socially constituted, and creatively enacted, representing the optimal realization of human potential within the complex ecology of person-earth-work-society relationships that define contemporary existence.

Based on our explorations, we can also identify four essential dimensions of realization that occur in holistic innovation:

1. Cognitive Realization: The understanding that transforms our perception and interpretation of reality. This includes the "aha moments" and profound insights that change how we see the world. Cognitive realization prepares the ground for all other forms of realization.

2. Technical Realization: The practical implementation of ideas into functioning protocols, products, services, or systems. This involves the engineering and development work that translates concepts into tangible reality. Technical realization requires expertise and attention to detail.

3. Social Realization: The collective adoption and integration of new possibilities into shared lifeworlds. This is about how innovations become part of our collective reality through consensus, adoption, and normalization. Social realization is inherently political and negotiated.

4. Inner Realization: The transformation of consciousness and character itself, both individual and collective. This is the deepest level of realization, where not just external reality but our very mode of experiencing reality is transformed. Inner realization is what makes us philosophically healthy, provided it takes into account our intercreation with other forms of life, natural and social.

These dimensions are not separate processes but interwoven aspects of the same endeavor. In truly holistic innovation, all four dimensions must be engaged. Technical realization without inner realization leads to soulless products; inner realization without technical and social manifestation may remain mere fantasy; cognitive realization without social realization creates recluse visionaries; social realization without cognitive understanding produces shallow trends.

Cognitive realization operates at the frontier where understanding meets creation. As Merleau-Ponty (2012 [1945]) observed in his *Phenomenology of Perception*, perception is not passive reception but active construction. When we realize something cognitively, we are not merely recognizing an external fact but participating in the possible creation of a meaningful pattern. This creative aspect of cognition has been consistently demonstrated in neurological research (Thompson, 2007; Varela et al., 1991), which shows that the brain does not simply record sensory data but actively constructs aspects of experiential reality through complex interpretative processes.

In my work with institutions or corporations, I observed this dimension of realization when engineers and managers experienced shifts in how they understood the systems they were a part of. For one senior researcher at Vattenfall R&D, the cognitive realization that electricity was not merely a commodity, but a social and ecological relationship potentially transformed her entire approach to technological innovation. As she explained: "I suddenly saw that what we call 'power' is not just electrons flowing through wires but a complex web of relationships between humans, technologies, and natural systems."

This cognitive transformation—from possibility to the intuition of a wider compossibility—illustrates what philosopher of science Thomas Kuhn (1962) called a "paradigm shift," not only new information within an existing framework but a fundamental reorganization of the framework itself. Such shifts cannot be forced or programmed but can be facilitated through practices that interrupt habitual patterns of thought and perception.

Cognitive realization often begins with what psychologist Daniel Kahneman (2011) called "System 2" thinking, slow, deliberate, conceptual cognition. It eventually has the capacity to transform our second nature, or "System 1" processes, the fast, intuitive, automatic responses that guide most of our daily interactions with reality. When a new understanding is fully realized (this takes many repetitive attempts), it may become part of our immediate, pre-reflective grasp of the world, changing not just what we think but how we think. But we are not there yet as a species: compossible thinking on a planetary scale is not yet our second nature. Perhaps the very idea of an autopilot eudynamia is a contradiction and we will always need to make a conscious effort, in other words to work to sustain holistic healthy systems.

Secondly, technical realization involves translating conceptual possibilities into material configurations and protocols that function in the physical world. This dimension engages directly with what Heidegger called the "ready-to-hand"; the practical, instrumental engagement with things that characterizes much of human activity (Heidegger, 1962). Technical realization requires not just theoretical knowledge but what anthropologist Tim Ingold (2013), inspired by the various traditions of meditation, contemplation, and craft, calls "knowing from the inside," the embodied, practical wisdom that comes from making and doing. That is where craft comes into play, with its necessary focus on detail. The craft of doing things well is one of the greatest human achievements. Contrary to what is usually claimed, there is a lot of thinking involved in craftsmanship and technical realization. Precise action is never mindless.

The process of technical realization necessarily involves negotiation with the material world, which offers both resistance and affordance to our creative intentions and inventions. As sculptor Michelangelo reportedly said of his work, "The sculpture is already complete within the marble block, before I start carving. It is already there, I just have to chisel away the superfluous material." This describes the dialogue between intention and material that characterizes all technical realization. Crealectics recognizes technology and method as a most admirable, if not sublime, feature or our engagement with the world. At the same time, automation is always dangerous to awareness.

Design pioneer Tim Brown argues that successful innovation requires what he calls "design thinking," a human-centered approach that integrates desirability, feasibility, and viability (Brown, 2009). A framework of compossibility by design provides a practical structure for the technical realization phase, ensuring that innovations not only function but address genuine human needs within viable organizational and economic contexts. Technical realization may take the form of prototypes, small-scale models, pilot projects, digital twins, and eventually full-scale implementations. Engineers must translate abstract concepts—for instance of renewable generation—into working systems that interact with existing infrastructure, natural environments, and human behavior. This translation is never straightforward but involves what philosopher of technology Don Ihde (1990) calls "technological intentionality," the way technologies mediate between human intentions and the world, transforming both in the process.

The success of technical realization depends on what computer scientist Herbert Simon (1996) called "the sciences of the artificial," the knowledge of how to design and implement human-made systems that interact effectively with the natural world. Technical realization requires a balance between analytical precision and creative synthesis, between scientific understanding and practical craft. The engineers of today need analytics, dialectics and crealectics.

Thirdly, Social realization involves the collective adoption and legitimation of new possibilities. This dimension acknowledges that reality is not merely physical but socially constructed through what sociologist Pierre Bourdieu (1977) called "habitus," the shared dispositions, practices, and understandings that constitute social life in a certain milieu. For an innovation to become socially real, it must be integrated into this collective habitus, becoming part of the taken-for-granted background against which social action unfolds. Alternatively, or simultaneously, the habitus must be evolved intentionally.

Simon Sinek's research into inspirational leadership reveals that the most successful innovations begin not with what they do or how they do it, but with why they exist, their fundamental purpose or underlying belief (Sinek, 2009). This insight is crucial for social realization, as innovations that clearly articulate their "why" create what Sinek calls "emotional resonance," one that facilitates adoption and integration into collective lifeworlds.

Foucault's analysis of power and knowledge reminds us that social realization is inherently political, involving negotiations over what counts as legitimate knowledge and practice (Foucault, 1980). Some possibilities are readily incorporated into existing power structures, while others face resistance precisely because they challenge these structures. As philosopher of science Sandra Harding (2006) has shown among many others, what counts as "real" or "objective" knowledge is also shaped by social positions and interests, making social realization a site of contestation and potential transformation.

At Vattenfall R&D, I witnessed or heard how technical innovations often faced resistance in implementation not because of technical inadequacies but because they challenged existing organizational hierarchies, professional identities, social dynamics, and aesthetic or cultural assumptions. Social realization requires what Latour (2007) calls "translation," the process by which diverse actors' interests are aligned around a new possibility, creating what he terms an "actor-network" that stabilizes and legitimizes the innovation. This involves not just rational persuasion but the cultivation of what anthropologist Clifford Geertz (1973) called "thick descriptions," rich narratives that make the innovation meaningful within existing cultural frameworks while potentially transforming those frameworks.

Last, but not least, inner realization involves the transformation of consciousness itself, our fundamental way of being in and experiencing the world. This dimension connects with what Maslow (1968) called "self-actualization" (or later self-transcendence), and what Ken Wilber (2000) terms "transpersonal development," the evolution of consciousness beyond conventional ego-identity toward more inclusive, complex, and integrated forms of awareness. This is where philosophical health practice complements crealectics. Innernovation is not merely a change in belief or understanding but what philosopher Pierre Hadot (1995) called a "spiritual exercise," a progressive transformation of one's way of being through practices that reshape perception, cognition, emotion, and identity. This transformation is both personal and collective, involving what psychologist Robert Kegan (1994) describes as shifts in the very "subject-object relationship" that structures consciousness, what we identify as "self" versus what we can observe and reflect upon.

In practicing the crealectic method as well as the SMILE_PH method (Sense-Making Interviews Looking at Elements of Philosophical Health), I observed how participants occasionally experienced profound realizations regarding their relationship to their work, their organization, and even their own identity. One manager described it as "seeing myself and my role from outside the usual frame, as if I had stepped back and could now see patterns I was usually too embedded in to notice." Inner realization relates to what psychologist Carl Rogers (1995 [1961]) called "becoming a person," the process of growing toward greater authenticity, a more virtuous character, an openness to experience, and authoritative authorship, a capacity for intercreative living. This shift in the quality of relationship extends not just to other humans but potentially to all beings, constituting what deep ecologist Arne Naess (1989) called an "ecological self" that identifies with the wider and compossible web of life.

Assessing the Four Evidences

Some questions remain. How do we evaluate the crealectic method and its results? How do we know when our practice is bearing fruit? How do we assess the quality of our creative realization? Here I propose four dimensions of evidence that correspond to the previously presented four dimensions of realization.

1. Cognitive Evidence: the clarity and coherence of our understanding, the "aha moments" that transform how we perceive reality. This manifests as conceptual breakthroughs, novel frameworks, unexpected connections between previously separate domains. When cognitive evidence flourishes, we experience not just new knowledge but transformed perception: seeing familiar realities with new eyes, with clarity and explicit precision.

2. Technical Evidence: the practical functionality of our innovations, their effectiveness in addressing real-world challenges. This includes not just whether something "works" in a narrow sense but how elegantly, efficiently, and appropriately it works within its context. Technical evidence concerns the embodiment of ideas in material configurations that function effectively in the physical world.

3. Social Evidence: the collective adoption and integration of new possibilities into shared lifeworlds, their impact on communities and cultures. This transcends mere market success to encompass how innovations transform relationships, institutions, and cultural patterns. Social evidence emerges when innovations become meaningful elements within collective narratives and practices.

4. Inner Evidence: the transformation of consciousness itself, the expansion of our capacity for creative participation. This most subtle dimension concerns how the process changes us as intercreators, enhancing our sensitivity, sophistication, and skill in engaging with possibility and compossiblity. Inner evidence manifests as increased creative confidence, expanded imaginative capacity, and deeper engagement with both constraints and possibilities.

These four dimensions can be further illuminated through a distinction between Creal Evidence and Real Evidence. This distinction doesn't create a binary opposition but rather highlights complementary aspects of creative evaluation. Let's start with what is most familiar in our mainstream world, the infamous Real Evidence of the evidenced-based ideology.

The Four Dimensions of Real Evidence

1. Repetitive – pertaining to recurrent and reproducible data, to consistency rather than novelty. This dimension values stability, reliability, and predictability, the capacity of an innovation for instance to produce the same results under similar conditions. Real evidence appears in the consistency of performance, in the reliability of function across contexts. At its best it represents what philosopher Whitehead (1978) called "permanence amid flux," the valuable stability that allows for everyday practice.

2. Neutral – based on universal rationality, on frameworks that claim perspective-independence. This dimension strives for what philosopher Thomas Nagel called "the view from nowhere," evaluation that transcends particular viewpoints. While complete neutrality is impossible (and perhaps undesirable), this dimension valuably reminds us to check our biases and expand our consideration beyond our immediate preferences. At its best, neutral evidence represents not absence of perspective but integration of multiple perspectives, what philosopher Hans-Georg Gadamer called a "fusion of horizons."

3. Measurable – pertaining to the physical or statistical realm, to what can be quantified and compared through standardized metrics. This dimension values precision, accuracy, and commensurability, the capacity to make meaningful comparisons across contexts through numerical representation. While measurement can become reductive, at its best it provides what philosopher Bruno Latour calls "immutable mobiles," stable references that can travel across contexts while maintaining their meaning, enabling coordination across difference.

4. Necessary – based on predictable causality, on the reliable connection between actions and outcomes. This dimension values the logical structure of reality, the dependable patterns that allow us to anticipate consequences and design accordingly. While necessity can become deterministic fatalism, at its best it represents what philosopher Charles Sanders Peirce called "the reality of laws," the reliable patterns that make scientific intervention possible.

In contradistinction, let's now try to define a new and complemenatary form of evidence, one that may not be easily accepted by hyper-realistic minds, but that is nevertheless very active in the world, even among scientists: Creal Evidence.

The 4 C's of Creal Evidence

1. Creative – pertaining to a process of growth and novelty, performative in nature. Creal evidence is not static but dynamic, not fixed but flowing. It concerns not just what has been created but the ongoing process of creation itself. We see this for instance in innovations that continue to evolve and generate new possibilities rather than becoming rigid and fixed. Consider how certain architectural works like Gaudí's Sagrada Familia embody an ongoing creative process, or how living technologies like permaculture gardens continue to grow and evolve rather than remaining static. Creative evidence manifests in the "generativity" of an innovation, its capacity to inspire further creation beyond its initial form.

2. Creditive – pertaining to belief, trust, and the cultivation of faith in possibility. This dimension addresses how innovations enhance or diminish our collective faith in what's possible. Do they expand our sense of the possible or contract it? Do they cultivate trust or erode it? In applying the crealectic method with groups, I observed how successful innovations build what sociologist Anthony Giddens (1991) calls "active trust," not blind faith but a dynamic confidence based on demonstrated integrity. Creditive evidence appears when innovations strengthen not just specific beliefs but the underlying capacity to believe in possibility or compossibility.

3. Constructed – reflecting a perspective and a structure that acknowledges its own participation in reality. While Real evidence often pretends to a "view from nowhere," Creal evidence explicitly acknowledges its viewpoint and constructive nature. It embraces what philosopher Donna Haraway calls "situated knowledges," recognizing that all evidence emerges from particular perspectives and contexts. This doesn't make it less valuable but perhaps more honest and potentially more integrative, capable of incorporating multiple perspectives rather than claiming false or premature universality. When innovations explicitly acknowledge their origins in particular needs, values, cultures and contexts, they demonstrate constructed evidence.

4. *Compossible* – pertaining to collective possibility rather than mere necessity, belonging to what should be shared rather than just what is or must be accepted. This dimension addresses how innovations relate to other possibilities, whether they enhance or diminish the field of potential at the crossroad of person, earth, work and society. Compossible evidence appears when innovations create what philosopher Ernst Bloch called "concrete utopia"—not fantasy disconnected from reality but real possibility that expands what's achievable. It's about verifying innovations that not only address immediate needs but create new shared spaces for further co-development of human and more-than-human beings.

True innovation engages all these dimensions simultaneously, creating evidence that is both Creal and Real, both qualitative and quantitative, both subjective and objective. The crealectic practice helps us navigate these dimensions more consciously, suggesting innovations that are not merely functional in a given world but meaningful across diverse lifeworlds, not just useful but beautiful, not only financially or technologically successful but holistically significant.

This integrated evaluation rejects the bifurcation of nature into primary (objective, measurable) and secondary (subjective, qualitative) qualities. Instead, it recognizes that reality itself is simultaneously Creal and Real, simultaneously flowing process and stabilized pattern. Our evaluation should reflect this ontological richness, engaging the full spectrum of evidence rather than artificially restricting ourselves to a single dimension, the tyranny of reductionism. The fact that there are few tools nowadays to evaluate the four aspects of Creal evidence does not mean that this dimension should be ignored as non-existent.

In working with engineers for instance, one would hope to observe how this multidimensional evaluation transforms their relationship with innovation. Rather than judging success solely by technical parameters or market metrics, engineers may need to consider how their work simultaneously transforms understanding (cognitive evidence), functions in practice (technical evidence), shapes communities (social evidence), and develops their own creative capacities (inner evidence). This more comprehensive evaluation will not diminish rigor but enhance it, creating more robust innovations that addressed needs across multiple dimensions.

As we cultivate crealectic practice, we develop greater sensitivity to all these dimensions of evidence, becoming capable of more nuanced and comprehensive evaluation. We learn to recognize not just whether something "works" in a narrow technical sense but whether it contributes to the eudynamia of lifeworlds across multiple scales and domains. This expanded evaluation is essential for addressing the complex challenges we face, which cannot be reduced to single dimensions or metrics without increased existential risk.

The Integration of Dimensions

These four dimensions of realization—cognitive, technical, social, and inner—form a holistic matrix of transformation that resonate with the four dimensions of reality as defined in the CIPHER model: person, earth, work, society. Each dimension interpenetrates and influences the others in various ways. For example, the inner realization of the person, facilitated by philosophical health, may catalyze new cognitive insights about life on earth, which in turn inform technical approaches embodied in ways of working, which then require societal legitimation, which may further transform inner consciousness, and so on in recursive spirals of intercreation (Fig. 6.3).

What makes the crealectic method distinctive is its explicit attention to the inner dimension of the person that is often neglected in conventional innovation processes, but also the compossible dimension of eudynamic lifeworlds. By integrating inner transformation with worldmaking, the crealectic method offers a more comprehensive approach to realization than methods that focus primarily on technical or utilitarian dimensions alone.

Realization is of course not only an individual matter; it is inherently collective. New technologies despite their supposed superiority are sometimes rejected by users for various reasons. The decision is democratic or at least needs a certain societal consensus. Realizing will always be a negotiation between different interests. This collectivization of the real has sometimes been called esprit de corps, by which united groups make things happen, a phenomenon I studied in my book *Ensemblance* (de Miranda, 2020b). A prime example is the Bourbaki Group, where young

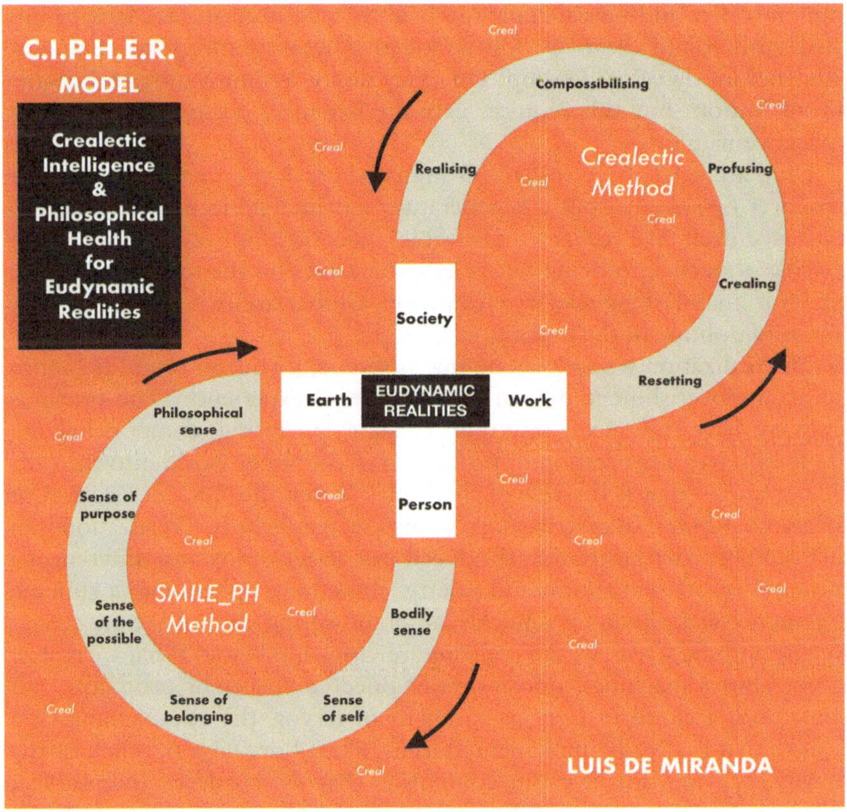

Fig. 6.3 The C.I.P.H.E.R. model framework

mathematicians in the early twentieth century realized the re-unification of mathematics by distributing their identity in a fictitious figure, Nicolas Bourbaki, an invented signature that represented the sum of all their individualities and aggregated skills. They displayed such esprit de corps and abnegation, such sense of collective purpose, that they managed to unify mathematics in a way no individual could have achieved.

We must recognize that there is often a discrepancy between how important we imagine our projects to be and how important they are perceived by the world. Usually, the world does not (immediately) see the importance we attribute to our realizations. Moreover, there are legal, administrative and bureaucratic obstacles to any innovation. Many promising embryos of creativity will be killed by real or imagined complications before they can develop. One may desire an innovation to become an institution, but this requires extreme patience, persistence, luck and collective will.

Ultimately, what we realize through the crealectic method is not just products or systems, but a regenerative and intersubjective approach to existence itself. We realize compossible lifeworlds: not only bringing our specific projects into being but placing ourselves on the meta-level where we try to manifest a planetary civilization. Global reality not as cacophony but as symphony in the making.

The realization we seek is one that is serene, joyful, and rich. It cannot be an impoverishment but an enrichment of experience. The previous steps we have traversed—resetting, crealing, profusing, compossibilizing are themselves a realization or projection of our inner world into the outside cosmos—that is the *ekto* of crealectics (From the Ancient Greek root meaning *outside*). In certain cases, what is seen as failure of realization is in fact the best part: nothing is produced but an air of joy and difference, a moment of contemplation and poetry, something ephemeral, a sublime emergence that disappears quickly, an empowering dream. We are realizing life and the Creal, we are realizing freedom and being. What we realize is not always machines, not always engineered spaces or protocols, but sometimes a cloud of hope, a moment of fervor, the reawakening of a feature that is already implemented in us, like a breathing of relief, or the gravity of living. But it is our contention that the crealectic method can also be useful in addressing very practical concerns, and it transcends the usual divide between dreamers and makers.

When was the last time we felt enchanted? When we joyfully access our inner world, but also when we feel adequation between our dreamworld and our real world.

At the end of the day, what we realize is a purpose, a higher purpose. What we realize is compossibility and eudynamia. Why is this important? Because the opposite of a world of possibility and compossibility is a world of contradiction and depletion. A planetary civilization will allow for the coexistence of many lifeworlds. A planetary disaster will destroy the very possibility of sharing a world.

Practical Approaches to Realization

Several practical approaches have proven effective for facilitating the translation from possibility into reality. These approaches engage multiple dimensions of realization simultaneously, creating synergies that enhance the overall process. Here are 10 examples among many others:

1. Prototyping: Creating small-scale, experimental versions of innovations that can be tested, refined, and scaled. Prototyping allows for learning through making and reduces the risk of large-scale implementation by providing adequate prior understanding. Donald Schön (1983) describes this as a "conversation with the materials of a situation," where each prototype or model reveals new aspects of the possibilities and compossibility being explored. For instance, prototyping energy management systems allows engineers to discover unanticipated interactions with existing infrastructure that would have been difficult to predict theoretically. The Älvkarleby facility of Vattenfall R&D, where I held several crealectic workshops, is known for its small-scale real-life testing models, for instance the "Laxelerator" (in Swedish, *lax* means salmon). The Laxelerator is a 30 meter long water pool in which different kinds of flowing water environments can be simulated, such that the behaviour of fish can be studied and predicted in fishway entrances for upstream and downstream migration in the context of hydropower facilities.

2. Iterative Development: Implementing innovations by breaking cycles of creation, testing, and refinement rather than in one comprehensive deployment. This approach, formalized in methodologies like Agile (Highsmith, 2002), acknowledges the emergent nature of realization; we cannot fully anticipate how a possibility will manifest until we begin to realize it, but we may be able to decompose compossibility into testable constituents. The iterative approach creates what complexity theorist Dave Snowden calls "safe-to-fail experiments" (Snowden & Boone, 2007), where each iteration provides learning without catastrophic risk.

3. Boundary Objects: Creating artifacts, models, or concepts that facilitate communication across different communities of practice. Sociologist Susan Leigh Star (1989) identified how certain objects can serve as translation devices between different social worlds, enabling collaboration without requiring consensus. For instance, visual models can function as boundary objects, allowing engineers, managers, community stakeholders, and policy makers to communicate across their different perspectives and expertise. When redeveloping a former industrial district, for instance, planners may create detailed 3D scale models and digital visualizations that serve as boundary objects across multiple communities of practice, architects, residents, city council members, environmental scientists. The CIPHER model is a boundary object.

4. *Participatory Implementation*: Involving diverse stakeholders in the realization process, ensuring that various perspectives and needs are addressed. This increases the likelihood of social realization while also creating opportunities for inner transformation. Political scientist Elinor Ostrom (1990), in her studies of common resource management, demonstrated how participatory approaches lead to more sustainable and effective implementations than top-down impositions. For instance, community energy projects that involve local residents in design and governance prove more resilient than those designed without such participation. Here mutual understanding can be obtained in depth by applying the SMILE_PH method (Sense-Making Interviews Looking at Elements of Philosophical Health).

5. *Narrative Construction*: Developing compelling stories about the innovation that help others understand its value and place in their lifeworlds. Narratives facilitate cognitive and social realization by connecting new possibilities with existing meanings and values. As cognitive scientist George Lakoff (2008) has shown, meaningful change requires not just new information but new frames, conceptual structures that organize how we interpret information. Powerful narratives provide these frames, making innovations comprehensible and meaningful. Narrative should not be seen as a communication tool and abandoned to advertising agencies, but it should be anchored in a well-defined grand purpose and corporate or institutional authoricity.

6. *Communities of Practice*: Fostering groups that share a concern or passion for something they do and learn how to do it better through regular interaction. In my book *Ensemblance*, I described how these communities transmit not just explicit knowledge but tacit understanding and identity: duty, drive, distinction and discipline become possible through a fifth D, a grand destination (de Miranda, 2020b). For instance, communities of practice and vocation (Weber, 1946) around renewable energy technologies become crucial vehicles for realizing innovations, as they facilitate the transfer of both technical knowledge and the values and perspectives that made the innovations meaningful. The Open Source Appropriate Technology community demonstrates how newcomers learn not just technical skills for building solar concentrators and wind turbines,

but also embody the values of energy democracy and community empowerment through legitimate peripheral participation—moving from basic assembly tasks to becoming practitioners who can improvise context-appropriate solutions using local materials.

7. Reflective Practice: Maintaining conscious awareness of how realization is unfolding, learning from successes and failures, and adjusting course accordingly. This supports all dimensions of realization by creating feedback loops between action and understanding. Organizational theorists Chris Argyris and Donald Schön (1978) distinguished between "single-loop learning" (adjusting actions to achieve desired outcomes) and "double-loop learning" (questioning the assumptions and values that define what outcomes are desired). Reflective practice enables both forms of learning, making realization more adaptive, transformative, and philosophically healthy.

8. Systemic Facilitation: Using skilled facilitation to guide groups through complex, multidimensional change processes. This approach, developed in traditions like Process Work (Mindell, 1995) and Theory U (Scharmer, 2009), acknowledges that realization involves not just rational planning but emotional, social, and consciousness dimensions that require skilled navigation. For instance, facilitated dialogues help groups move through the discomfort and uncertainty that often accompany significant innovation, preventing premature closure or regression to familiar patterns. The CIPHER model allows for alternating between personal dialogue and collective creativity, ideally using the two methods with the same group of people.

9. Living Laboratories: Creating real-world environments where innovations can be tested and refined in authentic contexts. This approach, pioneered in settings like MIT's Media Lab (Brand, 1987), recognizes that many aspects of realization only become apparent in the complex interactions of actual use. Living laboratories engage all four dimensions of realization simultaneously, allowing for integral development rather than piecemeal implementation. They are also spaces of interdisciplinary exchange between researchers, citizens, the government, nature and industry (users, providers, enablers, consumers, inhabitants, etc).

10. Gravity Itself: Humanity is a time accelerator, tending to do everything faster than evolution itself. Shortcuts are certainly a feature of a creative mind, in some cases, but so is the road less taken, which is not always the shortest one. The most difficult quality of realization is patience, the patience to be a strange attractor, a gravity core of purpose that slowly—very slowly—bends reality by attracting what it admires. In Latin, the adage said *Festina lente*: "Make haste slowly." In working with large multinational corporations like Vattenfall, one should not expect linear progress, nor fast-paced transformation, and there is wisdom in being both a pioneer and a tortoise rather than a hare.

Ethical Considerations in Realization

Probably more than the previous stages of the crealectic method, the process of realization raises ethical questions that must be addressed for holistic innovation to contribute positively to human eudynamia and ecological flourishing. Philosopher Hans Jonas (1984) proposed that modern technology has created an "imperative of responsibility" that requires us to consider the long-term consequences of our innovations upon our lifeworlds. This imperative is especially relevant to the realization phase, where possibilities begin to have concrete effects when implemented.

At least four ethical principles may guide responsible realization:

1. Precautionary Principle: When an innovation poses potential threats of serious or irreversible harm, lack of full scientific certainty should not be used as a reason to postpone measures to prevent that harm. This principle, articulated in environmental ethics (Raffensperger & Tickner, 1999), acknowledges the inherent uncertainty in realization processes and prioritizes caution when stakes are high.

2. Distributional Justice: The benefits and burdens of innovation should be distributed fairly, with special attention to impacts on vulnerable populations. Environmental justice scholar Robert Bullard (2000) has documented how the negative consequences of technological change often fall disproportionately on marginalized communities, while benefits accrue to the already privileged. Ethical realization requires deliberate attention to these distributional effects. For example, the development of large-scale wind farms in Nordic regions, among others by Vattenfall, has created a conflict where the benefits of renewable energy largely flow to European markets and urban populations, while the environmental and cultural costs may be borne disproportionately by Indigenous Sami communities whose traditional reindeer herding practices are disrupted by turbine installations that fragment migration routes and reduce grazing lands.

3. Procedural Justice: Those affected by an innovation should have meaningful opportunities to participate in decisions about its realization.

Political philosopher Iris Marion Young (2000) argues that legitimate decisions require not just fair outcomes but fair processes that include all relevant perspectives. Participatory approaches to realization enhance both the ethical legitimacy and the effectiveness of innovations. For example, Vattenfall modified its Storlandet wind farm design after facing strong criticism from the Gällivare forest Sámi village, illustrating how meaningful participation can lead to project changes—though such modifications came only after initial plans had already been developed without adequate Indigenous consultation.

4. Intergenerational Responsibility: Realization should consider impacts on future generations, not just present constituencies. Philosopher Edith Brown Weiss (1989) describes this as a principle of "intergenerational equity" or, as it is called in psychology, *generativity*, that requires us to preserve options, quality, and access for future persons. This principle of social generativity is especially relevant for energy innovations with long-term ecological consequences. For example, Vattenfall's WIN@sea initiative demonstrates generativity by developing offshore wind farms that not only generate renewable energy but also cultivate seaweed that helps regenerate marine ecosystems for future generations.

These various ethical considerations are not external constraints on realization but integral aspects of holistic innovation. As Nussbaum (2011) argues, ethical reasoning is not separate from practical reasoning but a form of it, a way of thinking about how to realize human capabilities and ecological flourishing. The crealectic method incorporates ethical reflection, not as paralysis by analysis, but throughout the realization process, recognizing that how we realize innovations is as important as what we realize. Thinking about the ethical and multi-stakeholder dimensions by design, long before the implementation, may avoid the costly modifications of reactive virtue, triggered by conflicts or incidents. This works better if all parties, especially those who systematically oppose change, are made to access philosophically healthy self-understanding even before they discuss the issues at stake.

These approaches should be tailored to the specific context and nature of the innovation. What works for energy systems for instance may not work for educational reforms or artistic endeavors. The art of realization lies in matching the approach to the unique characteristics of each innovation and its intended lifeworlds (de Miranda, 2020a). The methodologies outlined above provide a toolkit, not a formula; resources to be adapted creatively to the challenges and opportunities of each realization process.

At the end of the day, there is ideally a relief in the release of realizing. What we may release in the world is a resonance between the aspirations of all beings, as if in a gigantic collective dream (Jung, 2014). And isn't that what our reality is anyway, a gigantic collective dream or nightmare? The crealectic method navigates between flow, blast, and gravity, recognizing that creativity operates sometimes as gentle stream, sometimes as eruption, often as pulsing nurturing rhythm that combines both. We learn to develop what we might call "blast-flow-gravity literacy"—the capacity to engage consciously with creation's full spectrum of intensities: cascades, pulses, attractions, whispers, detonations, drifts, eruptions, fluxes, turbulences, streams, coherences, caresses and sparks...

Potis in Latin is related to potential and means to be a master, to be powerful, to be able to. Realization is about empowering future generations via practices and affordances. As we are about to conclude our exploration of the crealectic method, we recognize that we have not merely described a technique but outlined a way of being in the world (de Miranda, 2021a), one that honors both the creative abundance of possibility and the harmonious integration of compossibility. It's idealistic, but it's also pragmatic. In the tension between the imaginal and the actual, between the virtual and the real, we discover our highest calling: to participate consciously in the ongoing intercreation of reality itself, to realize not just products or services but shared worlds of meaning and value—a planetary civilization.

The journey doesn't end with realization in society but circles back to the embodied persons that we are, as we continually renew our creative engagement with life via our bodily sense, sense of self, sense of belonging, sense of the possible, sense of purpose and the philosophical sense that allows us to work not against the Earth but inspired by it. The CIPHER spiral continues, Crealectic Intelligence and Philosophical Health for Eudynamic Realities, inviting us to ever deeper participation in the creative unfolding of existence. In this infinite loop lies the promise not just of innovation but of sustained and distributed possibilities for regenerative growth. The Anthropocene is dying, long live the Regenerocene!

References

Argyris, C., & Schön, D. A. (1978). *Organizational learning: A theory of action perspective*. Addison-Wesley.

Boltanski, L., & Chiapello, E. (2005). *The new spirit of capitalism*. Verso.

Bourdieu, P. (1977). *Outline of a theory of practice* (R. Nice, Trans.). Cambridge University Press.

Brand, S. (1987). *The Media Lab: Inventing the future at MIT*. Viking.

Brown, T. (2009). *Change by design: How design thinking transforms organizations and inspires innovation*. HarperBusiness.

Brown Weiss, E. (1989). *In fairness to future generations: International law, common patrimony, and intergenerational equity*. United Nations University.

Buber, M. (1970). *I and Thou* (W. Kaufmann, Trans.). Charles Scribner's Sons. (Original work published 1923).

Bullard, R. D. (2000). *Dumping in Dixie: Race, class, and environmental quality* (3rd ed.). Westview Press.

Damasio, A. (1999). *The feeling of what happens: Body and emotion in the making of consciousness*. Harcourt Brace.

de Miranda, L. (2020a). *Paridaiza*. Snuggly Books.

de Miranda, L. (2020b). *Ensemblance: A Transnational Genealogy of Esprit de Corps*. Edinburgh University Press.

de Miranda, L. (2021a). Crealectic intelligence. In V. Glăveanu (Ed.), *The Palgrave encyclopedia of the possible* (pp. 1–8). Palgrave Macmillan.

de Miranda, L. (2021b). The healing-growth future of humanity: Regenerative politics and crealectic care. *Acta Academica, 53*(2), 143–163.

de Miranda, L. (2024). *Philosophical health: A practical introduction*. Bloomsbury Publishing.

Dewey, J. (2005). *Art as experience*. Perigee Books. (Original work published 1934).

Diamond, J. (1997). *Guns, germs, and steel: The fates of human societies*. W.W. Norton & Company.

Dietrich, A. (2004). The cognitive neuroscience of creativity. *Psychonomic Bulletin & Review, 11*(6), 1011–1026.

Foucault, M. (1980). *Power/knowledge: Selected interviews and other writings, 1972–1977* (C. Gordon, Ed.; C. Gordon, L. Marshall, J. Mepham, & K. Soper, Trans.). Pantheon Books.

Geertz, C. (1973). *The interpretation of cultures*. Basic Books.

Giddens, A. (1991). *Modernity and self-identity: Self and society in the late modern age*. Stanford University Press.

Gilbert, D. T., & Wilson, T. D. (2007). Prospection: Experiencing the future. *Science, 317*(5843), 1351–1354.

Hadot, P. (1995). *Philosophy as a way of life: Spiritual exercises from Socrates to Foucault* (M. Chase, Trans.). Blackwell.

Harding, S. (2006). *Science and social inequality: Feminist and postcolonial issues*. University of Illinois Press.

Harman, G. (2018). *Object-oriented ontology: A new theory of everything*. Pelican Books.

Heidegger, M. (1962). *Being and time* (J. Macquarrie & E. Robinson, Trans.). Harper & Row. (Original work published 1927).

Highsmith, J. (2002). *Agile software development ecosystems*. Addison-Wesley.

Ihde, D. (1990). *Technology and the lifeworld: From garden to earth*. Indiana University Press.

Ingold, T. (2013). *Making: Anthropology, archaeology, art and architecture*. Routledge.

Jonas, H. (1984). *The imperative of responsibility: In search of an ethics for the technological age*. University of Chicago Press.

Jung, C. G. (2014). *The archetypes and the collective unconscious* (R. F. C. Hull, Trans.). Princeton University Press. (Original work published 1959).

Kahneman, D. (2011). *Thinking, fast and slow*. Farrar, Straus and Giroux.

Kegan, R. (1994). *In over our heads: The mental demands of modern life*. Harvard University Press.

Kuhn, T. S. (1962). *The structure of scientific revolutions*. University of Chicago Press.

Lakoff, G. (2008). *The political mind: Why you can't understand 21st-century politics with an 18th-century brain*. Viking.

Latour, B. (2007). *Reassembling the social: An introduction to actor-network-theory*. Oxford University Press.

Maslow, A. H. (1968). *Toward a psychology of being* (2nd ed.). Van Nostrand Reinhold.

Maturana, H. R., & Varela, F. J. (1980). *Autopoiesis and cognition: The realization of the living*. D. Reidel Publishing Company.

May, R. (1975). *The courage to create*. W. W. Norton & Company.

Meadows, D. H. (2008). *Thinking in systems: A primer*. Chelsea Green Publishing.

Merleau-Ponty, M. (2012). *Phenomenology of perception* (D. A. Landes, Trans.). Routledge. (Original work published 1945).

Mindell, A. (1995). *Sitting in the fire: Large group transformation using conflict and diversity*. Lao Tse Press.

Naess, A. (1989). *Ecology, community and lifestyle: Outline of an ecosophy* (D. Rothenberg, Trans.). Cambridge University Press.

Nussbaum, M. C. (2011). *Creating capabilities: The human development approach*. Harvard University Press.

Ostrom, E. (1990). *Governing the commons: The evolution of institutions for collective action*. Cambridge University Press.

Penrose, R. (1989). *The emperor's new mind: Concerning computers, minds, and the laws of physics*. Oxford University Press.

Raffensperger, C., & Tickner, J. (Eds.). (1999). *Protecting public health and the environment: Implementing the precautionary principle*. Island Press.

Rogers, C. R. (1995). *On becoming a person: A therapist's view of psychotherapy*. Houghton Mifflin. (Original work published 1961).

Scharmer, C. O. (2009). *Theory U: Leading from the future as it emerges*. Berrett-Koehler Publishers.

Schön, D. A. (1983). *The reflective practitioner: How professionals think in action*. Basic Books.

Simon, H. A. (1996). *The sciences of the artificial* (3rd ed.). MIT Press.

Sinek, S. (2009). *Start with why: How great leaders inspire everyone to take action*. Portfolio.

Snowden, D. J., & Boone, M. E. (2007). A leader's framework for decision making. *Harvard Business Review, 85*(11), 68–76.

Star, S. L. (1989). The structure of ill-structured solutions: Boundary objects and heterogeneous distributed problem solving. In L. Gasser & M. N. Huhns (Eds.), *Distributed artificial intelligence* (Vol. 2, pp. 37–54). Morgan Kaufmann.

Taneja, H. (2022). *Intended consequences: How to build market-leading companies with responsible innovation*. Harvard Business Review Press.

Thagard, P. (2002). *Coherence in thought and action*. MIT Press.

Thompson, E. (2007). *Mind in life: Biology, phenomenology, and the sciences of mind*. Harvard University Press.

Thompson, L. (2013). *Creative conspiracy: The new rules of breakthrough collaboration*. Harvard Business Review Press.

Varela, F. J., Thompson, E., & Rosch, E. (1991). *The embodied mind: Cognitive science and human experience*. MIT Press.

Weber, M. (1946). Science as a vocation. In H. H. Gerth & C. Wright Mills (Eds.), *From Max Weber: Essays in sociology* (pp. 129–156). Oxford University Press.

Whitehead, A. N. (1978). *Process and reality* (Corrected ed.). Free Press. (Original work published 1929).

Wilber, K. (2000). *Integral psychology: Consciousness, spirit, psychology, therapy*. Shambhala.

Young, I. M. (2000). *Inclusion and democracy*. Oxford University Press.

CHAPTER 7

Conclusion: The Pluriverse of Compossible Worlds

Abstract The concluding chapter synthesizes the crealectic method within broader frameworks of consciousness evolution and planetary regeneration. It explores the transition from "anthrobotic" to "regenerative" modes of innovation, emphasizing the cultivation of what the author terms "planetary thinking" for long-term transformation. The chapter positions crealectics within a pluriversal worldview that honors diversity while seeking creative harmony across different lifeworlds and scales of existence.

Keywords Pluriverse • Regeneration • Consciousness evolution • Cathedral thinking • Anthrobot • Regenerocene • Long-term thinking • Planetary innovation • Transformation • Creative evolution

We have traversed the ocean of possibility, navigated the shores of compossibility, and return transformed. We stand at the edge of a new beginning. The journey through the five phases of the crealectic method—resetting, crealing, profusing, compossibilizing, and realizing—has brought us not to an end but to a threshold of ongoing creation. What remains is to contemplate what this journey will mean in the future and how it might transform our approach to innovation, consciousness, and the shared lifeworlds we inhabit (Fig. 7.1).

L. de Miranda, *Crealectics as a Creative Method*,
Palgrave Advances in Possibility Studies,
https://doi.org/10.1007/978-3-032-00427-7_7

Fig. 7.1 The creal principle for a planetary civilization, based on a wall painting by Luis de Miranda

Our voyage through the crealectic method has been not merely an exploration of techniques but an invitation to a fundamentally different relationship with creativity itself. As we conclude, let us not merely summarize where we have been but envision where we might go next, beyond method and into the lived experience of crealectic intelligence.

The Hidden Logic of Impossible Innovation

Our most persistent innovation blockages stem not from technical limitations but from our adherence to what mathematicians call "crisp logic," the analytic binary thinking that demands clear yes/no answers in a world of irreducible uncertainty. Masatoshi Sakawa's groundbreaking work on fuzzy sets reveals that the most complex optimization problems require what he terms "interactive multiobjective optimization," approaches that can work productively with multiple, potentially conflicting objectives simultaneously rather than reducing them to single metrics (Sakawa, 1993). He insists that the main characteristics of real-world decision-making are multidimensional and have multiple objectives including economic, environmental, social, personal and technical ones.

This mathematical insight illuminates a crucial limitation in conventional metric-based protocols. When we demand that ideas be instantly-practical, measurable, or immediately non-contradictory, we eliminate precisely those possibilities that might lead to breakthrough innovations. Malgorzata Mycielska's illustrated book on "impossible inventions" demonstrates this principle with humor: her quirky descriptions of strange breakthrough innovations reveals that most began as ideas that violated the logical or social constraints of their time (Mycielska et al., 2018). What we might call "Unimpossible thinking" operates by expanding the space of possibility rather than optimizing within existing constraints.

The crealectic method now reveals its deeper logic. By encouraging the generation of ideas that may initially seem impossible or contradictory, we first create what Sakawa would recognize as a "fuzzy solution space," a field of possibilities where future realities exist in degrees of potential rather than data for analytic projection.

Anthropologist Lewis Hyde's investigation into trickster figures across human cultures reveals that the trickster embodies a fundamental creative principle: the capacity to work at the boundaries between order and chaos, creating new possibilities by playfully violating established categories (Hyde, 1998). But Hyde's analysis goes deeper than folklore. The trickster, he argues, operates according to what he calls "crossroads thinking," the ability to hold multiple perspectives simultaneously without premature synthesis. The trickster is the divine joker of exchange, of crossroads, of hinges, of places where information or ideas from different groups intermingle. A eudynamic trickster inspires our CIPHER innernovation, at the crossroad of person, earth, work and society, but he works together with his more serious sibling the Compossibilizer, the spirit of worldmaking.

This transforms our understanding of the compossibilizing phase: rather than merely combining equivalent or comparable ideas, we engage in creative recomposition across domains, disciplines, and traditions, allowing possibilities to cross-pollinate in ways that no single perspective could orchestrate. We operate at the crossroads, liminal spaces where different worlds or logics may meet and new combinations become possible. The crealectic method activates such crossroads by bringing together diverse perspectives within a single creative process, enabling "boundary crossing" that generates genuinely novel possibilities in collective authorship.

The Fluctuating Nature of Wonder

But why does this boundary crossing work? Rediscovering the core intuition of philosophy itself—and Descartes's suggestion that admiration is the mother of all emotions, we may say that wonder operates as a mode of consciousness that allows us to perceive intricate connections that analytical thinking typically filters out. Admiration, surprise, and wonder are not merely an emotional response but an active form of ecological engagement that enables us to participate in the universe's own creative process.

This insight connects with Carl Jung's investigation of synchronicity, meaningful coincidences that suggest an acausal connecting principle operating alongside causality (Jung, 1960). Jung's work, now rediscovered by developments in quantum consciousness research, suggests that thinking itself may at least partly rely on quantum principles, where multiple possibilities exist in superposition until observation collapses them into specific actualities or perspectives. But thinking does not only happen in the brain; as we have argued in this book, thinking is an extended (earthly), embodied (personal), enactive (work-situated) and embedded (social) activity, as suggested by the 4E cognition research field in cognitive science (Newen et al., 2018).

The crealectic method requires a form of receptive attention rather than aggressive problem-solving. Wonder creates the cognitive conditions necessary for cosmic consciousness to operate, allowing us to perceive and work with (com)possibilities that reflect the various dimensions of our lifeworlds.

Physicist Giorgio Parisi's Nobel Prize-winning work on complex systems provides a scientific framework for understanding how emergent properties might connect with larger organizational processes in nature (Parisi, 2021). Parisi's research on "frustrated systems" reveals a crucial principle: nature consistently discovers what he demonstrates as stable configurations within complex solution landscapes, that is equilibrium states that preserve the capacity for further adaptive reorganization. Nature inspires us to become eudynamic.

Parisi's work illuminates why conventional innovation methods often fail. When we optimize for specific linear metrics or solve problems in isolation, we create local success that prevents global coherence. The crealectic method addresses this by working at multiple scales simultaneously: resetting individual consciousness while crealing collective fields of possibility, profusing diverse ideas while compossibilizing them into coherent wholes. At the same time, we don't want to close the door to a collaboration between quantitative approaches and qualitative ideation, and this is why the CIPHER model focuses on the four scales of Person, Earth, Work and Society (PEWS), an ensemble that is large enough to be comprehensive but perhaps specified enough such that some mathematical modelling might be one day added to the innovative process.

This multiscale approach reflects what physicist Brian Swimme and cultural historian Thomas Berry describe as the "universe story," a recognition that human creativity participates in cosmic and earthly creativity, and vice versa (Swimme & Berry, 1992). From this perspective, innovation becomes not just human problem-solving but conscious participation in the universe's own process of creative self-discovery. The crealectic method, understood from this holistic perspective, becomes a practice of planetary thinking, innovation that serves not just human needs but the creative evolution of the entire Earth system. This doesn't diminish human agency but expands it and relativizes it by recognizing that innovation operates within and as part of cosmic creativity.

Toward a New Understanding of Reality

What emerges from this integration of fuzzy logic, playful consciousness, quantum awareness, and cosmic creativity is a fundamentally new understanding of innovation. Instead of treating innovation as the application of human intelligence to passive matter, we see it as participation in the universe's own creative process, a process that operates through principles that transcend the limitations of purely analytical (data-based) or dialectical thinking (oppositional).

This understanding transforms our relationship to the challenges we face. Climate change, social inequality, technological disruption—these are not merely problems to be solved but "system states" that call for meta-solutions operating at multiple scales simultaneously and nurturing a sense of the possible and the compossible in all living beings. The crealectic method offers: an appraised device for this work.

The most profound innovations often emerge not from direct assault on problems but from playful and philosophical engagement with the conditions that generate the way we consider or define the problems. A crealectician participates in what mystics describe as "the great work" of conscious evolution. The crealectic method is more than an innovation protocol. It becomes a practice of conscious participation in the creative unfolding of reality itself, what we might call the universe's own process of learning to *innernovate*. In learning to create more holistically, we discover that we are not separate from creation but expressions of its deepest principles.

This recognition transforms not just what we create but who we understand ourselves to be as creators. We are not isolated problem-solvers imposing solutions on a reluctant world but conscious participants in lifeworlds that are continuously creating themselves through us and as us. The crealectic method offers a way to participate more consciously in this worldly compossibilization, and the ultimate horizon of this participation extends into the mystery of existence itself.

At Vattenfall R&D and in other organizational or individual contexts, I have seen how the crealectic method can inspire new approaches to complex challenges, even if participants do not immediately perceive what is going on. Where conventional approaches often prioritize efficiency, control, and predictable outcomes, crealectics embraces personal and collective responsibility. It recognizes that the most profound innovations often emerge from conscious and reflective participation in the creative flow of reality itself. Embracing the unpredictable and the unheard-of is compatible with running corporations, institutions and other pragmatic projects. This is a daily combat, as discouraged routines, furious bureaucracies, and coward surrendering to the status quo are all too often presented to us for breakfast, lunch, and dinner.

The crealectic method cultivates an intersubjective relationship with creation, measuring its success not just by what it produces but by how it slowly transforms those who practice it, provided it is repeated faithfully. Does it enhance their senses of the possible and the compossible? Does it foster more harmonious relationships between diverse elements of reality? Does it contribute to the realization of philosophically healthy lifeworlds? These questions point toward qualitative rather than merely quantitative evaluation.

The noonaut—the explorer of the mind we met in the introduction of this book—travels not only outward but inward and between worlds, exploring the myriad possibilities that exist in the multiversal mind. This is not escapism but a profound engagement with reality itself, for as we've seen throughout these chapters, what we call "reality" is not a fixed substance but a continuous creative process in which we participate.

The crealectic method offers navigation instruments for this voyage. Like the *cogito* of Descartes, it represents a fundamental reorientation of consciousness, a shift from seeing ourselves as passive recipients of a pre-given reality to active participants in its continuing creation.

Creo, ergo sum. I create, therefore I am.

This more-than-human rhythm, this intercreative dance with time, is what the crealectic method helps us navigate more consciously. It resists what sociologist Hartmut Rosa (2015) calls our "desynchronization" from natural and social rhythms in modernity, our acceleration beyond the tempos that allow for genuine creative participation. The method's five phases restore a more organic rhythm, a conscious synchronization with the creative pulse of a rewilded yet harmonious reality.

From Anthrobot to Regenerocene

The predominant mode of contemporary innovation is "anthrobotic," a hybrid of human hybris and mechanical execution that often enhances productivity at the expense of authenticity, real creation and meaning. The Anthrobot, this blending of machine and humanity, excels at analytics, at dissecting reality into manipulable fragments. It is skilled at dialectics, at resolving contradictions through binary and Manichaean synthesis. But it struggles with crealectics, with the fluid intercreation that characterizes our deep transformations.

What would it mean to transcend the Anthrobot? Not to abandon our technological capacities but to reintegrate them within a more expansive consciousness, one that recognizes technology as an extension amon others of our creative participation rather than a substitute for it. The Anthrobot is the child of the Anthropocene. The crealectic method points toward the "Regenerocene," an era where human and non-human life flourish through mutual regeneration and intercreation, an epoch where future generations are included in our practice of eudynamia, the cultivation of a good relationship with the possible and the compossible.

The Regenerocene replaces extraction with composition, competition with symbiosis, control with intercreation. It would recognize what Margulis (1998) demonstrated about evolution: that symbiosis and cooperation have been more important than competition in shaping life's development.

My crealectic work at Vattenfall R&D—which I offered as the real and metaphorical case study along this book—offers a glimpse of this potential transformation. An energy company represents in microcosm the central challenge of our time: how to generate and distribute *power* in all its fruitful forms and in ways that enhance rather than diminish life's creative potential. The crealectic workshops there explored—and are still exploring—possibilities beyond conventional frameworks, not just more efficient technologies but fundamentally different relationships between energy, community, and natural systems.

As one engineer remarked after a workshop: "I've spent my career thinking about how to extract energy more efficiently. Now I'm beginning to think about how to participate in energy's flows, as a whole process of which we are a component among others." This shift from extraction to participation, from control to intercreation, exemplifies the potential of crealectic thinking to transform not just what we create but how we create.

As I explained in the introduction and beyond, the crealectic method exists within a larger framework,the CIPHER model: Crealectic Intelligence and Philosophical Health for Eudynamic Realities. This model recognizes that creative intelligence requires a foundation of philosophical coherence, a harmony between our values, meanings, and actions. Without this foundation, creativity risks becoming disconnected from purpose, innovation detached from shared meaning.

The word "cipher" carries multiple meanings relevant to our discussion. In cryptography, a cipher is a method of transforming a message to conceal or reveal its meaning. In this sense, CIPHER helps us decode the hidden potentials within everyday reality, revealing the extraordinary within the ordinary at the crossroads of personhood, earthly belonging, ways of working and social embedding. But cipher also means in English "zero," the empty circle that symbolizes both nothingness and infinite potential. In this sense, CIPHER represents the practiced emptiness, the ultimate possibility from which genuine creation emerges.

The crealectic method addresses the "CI" of CIPHER, the Crealectic Intelligence and eupoietic wellbeing that allows us to participate more consciously in reality's creative unfolding. But it remains incomplete without the "PH," the Philosophical Health and eudaimonic wellbeing that gives this participation coherence, deep understanding and direction (de Miranda, 2024). Together, they generate "ER," Eudynamic Realities, experiences that expand and harmonize rather than contract our sense of the possible. This integration of creative process and philosophical coherence at the crossroad of person, earth, work and society distinguishes the crealectic approach from many innovation methodologies. It recognizes that how we create is inseparable from who we are and what we value. As German philosopher Martin Heidegger (1977) observed, technology, protocols and methods are not merely a means but can become a mode of revealing, a way of bringing-forth that expresses our fundamental relationship with being itself.

The five phases of the crealectic method are not merely sequential steps but recurring moments in a spiral of creative engagement. We reset not once but continually, clearing space for fresh perception. We creal not occasionally but constantly, reconnected with the creative flow that sustains us. We profuse, compossibilize, and realize in ongoing cycles, each cycle transforming not just our creations but ourselves as creators.

This ongoing practice develops what psychologist Ellen Langer (2014) calls "mindful creativity," a heightened awareness of the creative potential in each moment, a capacity to perceive and engage with the present beyond habitual patterns. It cultivates what philosopher Gilles Deleuze (1994) termed nomadic versus dogmatic thought, thinking that moves freely and with fluidity across territories rather than remaining fixed within predetermined boundaries. But most fundamentally, crealectic practice develops what the Greek philosophers called *phronesis*—practical wisdom—the capacity to discern and act in accordance with what truly enhances life, balancing knowledge, ethical sensitivity, disciplined craft and lived experience in pursuit of eudaimonic wellbeing. This is embodied and enactive understanding, not a theoretical construction of the ivory tower kind. Crealectic philosophy is a lived participation in the creative unfolding of reality itself.

The Pluriverse of Compossible Worlds

One of the horizons of crealectic practice is what James (1996) called the "pluriverse," a reality composed of multiple lifeworlds, each with its own integrity yet capable of creative interaction with others. The crealectic method helps us navigate this pluriverse by discovering patterns of potential harmony and dwelling amid difference.

Transition town movement founder Rob Hopkins argues that many of our current crises stem from what he terms "a failure of imagination," our relative incapacity to envision alternatives to prevailing destructive patterns (Hopkins, 2019). His work on community-scale transformation demonstrates how the systematic cultivation of imagination can catalyze practical innovations that regenerate both social and ecological systems. This navigation requires what philosopher Hans-Georg Gadamer (2004) called a "fusion of horizons," an expansion of perspective that allows us to perceive connections across apparent divides. It involves, as we have seen, what Hofstadter (2007) described as "strange loops," the recursive patterns that connect different levels of reality into meaningful wholes. These capacities don't eliminate the challenges of innovation but transform our relationship to them, making them opportunities for creative engagement rather than obstacles to be overcome.

The German philosopher Peter Sloterdijk (2011) speaks of "plural spherology," the recognition that we inhabit not a single encompassing sphere but multiple interlinked "foam" structures of phenomena, lifeworlds of meaning and existence, each with its own atmosphere and logic. The crealectic practice helps us navigate these plural spheres, not by imposing a dogma but by discovering pathways and compossiblities between them, creating what Sloterdijk calls "air bridges" that allow for movement and exchange without collapsing diversity. Nevertheless, the eudynamic purpose of the CIPHER model recognizes that if we are to build a planetary civilization we need a shared cosmology beyond the romanticism of alterity and the disorders of communitarianism. In compossibilizing our global existence, some pruning will be necessary. Nihilistic attitudes need to be identified and diplomatically transformed into life-affirming practices. As Voltaire anticipated in his *Candide*, crealecticians are gardeners. As Leibniz professed in his *Monadology*, not every possible may coexist in the garden we call Earth.

The pluriversal perspective challenges the modern project of reduction and unification through mere metric abstraction, what Latour (1993) calls the "work of purification" that separates nature from culture, subject from object, science from politics. Instead, crealectics embraces what a work of translation, the creation of networks and hybrids that allow for symbiosis across differences without erasing diversity. The crealectic method contributes to this work of intercreative transmutation, helping us create what anthropologist Arturo Escobar (2018) calls "designs for the pluriverse," innovations that enhance rather than diminish the diversity of lifeworlds. This is not relativism but relationality, not the abandonment of shared truth but the recognition that truth itself emerges through creative engagement with difference.

As we reach the end of this book, we find ourselves at a new beginning, a threshold of continuing creation. The crealectic method is not a closed system but an open invitation, not a fixed map but a dynamic compass that orients us within the ever-shifting territories of possibility and compossibility.

In my work and life, I've witnessed how crealectic intelligence can extend the limits of what we think is achievable. I've seen engineers reconnect with the playful wonder that first drew them to their field, executives rediscover deeper meaning beneath data, teams transcend both individualism and groupthink to find new patterns of collaborative creation. I have also sometimes failed momentarily and seen how one step forward can become two steps back in working with companies or institutions. Sometimes I was tempted to forget my care for the eudynamic balance between person, earth, work and society—to live and let die. But a crealectician cannot give up because crealectics is the movement of life itself. The key to a patient path to transformation is to multiply the modes of actualization of our higher purpose. Our higher purpose should be eudynamia for all forms of life, as a result of eudaimonic and eupoietic forms of practice. Such is, I believe, the foundation upon which a planetary civilization might emerge.

Intercreating a planetary civilization will take time. Cultural philosopher Roman Krznaric's investigation into long-term thinking reveals how short-termism has become "the defining challenge of our time," constraining our ability to create innovations that truly serve future generations (Krznaric, 2020). The crealectic method addresses this challenge by cultivating what Krznaric calls "cathedral thinking"—the capacity to initiate projects whose full realization may extend far beyond our individual lifespans, thus honoring our responsibility to be, in his words, "good ancestors," stewards of a future we will only inhabit in spirit.

The noonaut's voyage is a continuous exploration of the ocean of compossibility, a recurring cycle that spirals ever outward and inward simultaneously. In the words of poet T.S. Eliot (1943): "We shall not cease from exploration, and the end of all our exploring will be to arrive where we started and know the place for the first time." The CIPHER loop regularly brings us back to where we started—the Creal itself—but with transformed awareness that allows us to participate more consciously in its unfolding. Eudynamia is an infinite loop, a Möbius strip on which we journey with awe and wonder.

Wonder is not just an emotional response but a cognitive skill that can be developed through practice (Davis, 2021), supporting the understanding that creative capacity can be systematically cultivated through appropriate methods. As you close this book and continue your voyage, may you carry with you not just techniques but perspective, not just methods but vision, not just knowledge but wisdom. May we discover ourselves anew as explorers of the endless ocean of mind, sailors on the sea of the compossible. May our voyage contribute to the creation of the best of all compossible worlds, not a perfect utopia but a living pluriverse where diverse lifeworlds flourish in eudynamia.

Allow me to invite you to engage with the crealectic method not as a producer, consumer or spectator, but as an intercreator, not as a follower but as a fellow explorer, leaping between lifeworlds while working, here and now, toward our planetary civilization. The crealectic journey is ours to navigate, ours to discover, ours to realize.

May the Creal be with you.

References

Davis, J. (2021). *Tracking wonder: Reclaiming a life of meaning and possibility in a world obsessed with productivity.* Sounds True.

de Miranda, L. (2024). *Philosophical health: A Practical introduction.* Bloomsbury.

Deleuze, G. (1994). *Difference and repetition* (P. Patton, Trans.). Columbia University Press.

Eliot, T. S. (1943). *Four quartets.* Harcourt.

Escobar, A. (2018). *Designs for the pluriverse: Radical interdependence, autonomy, and the making of worlds.* Duke University Press.

Gadamer, H-G. (2004). *Truth and method* (J. Weinsheimer & D. G. Marshall, Trans.). Continuum. (Original work published 1960).

Heidegger, M. (1977). The question concerning technology. In *The question concerning technology and other essays* (W. Lovitt, Trans., pp. 3–35). Garland. (Original work published 1954).

Hofstadter, D. (2007). *I am a strange loop.* Basic Books.

Hopkins, R. (2019). *From what is to what if: Unleashing the power of imagination to create the future we want.* Chelsea Green Publishing.

Hyde, Lewis. (1998). *Trickster makes this world: Mischief, myth, and art.* Farrar, Straus and Giroux.

James, W. (1996). *A pluralistic universe.* University of Nebraska Press. (Original work published 1909).

Jung, C. G. (1960). *Synchronicity: An acausal connecting principle.* Princeton University Press.

Krznaric, R. (2020). *The good ancestor: A radical prescription for long-term thinking.* The Experiment.

Langer, E. J. (2014). *Mindfulness* (25th anniversary ed.). Da Capo Press.

Latour, B. (1993). *We have never been modern* (C. Porter, Trans.). Harvard University Press.

Margulis, L. (1998). *Symbiotic planet: A new look at evolution.* Basic Books.

Mycielska, M., Mizieliński, M., & Mizieliński, D. (2018). *Impossible inventions: Ideas that shouldn't work.* Gecko Press.

Newen, A., De Bruin, L., & Gallagher, S., eds. (2018). *The Oxford handbook of 4E cognition.* Oxford University Press.

Rosa, H. (2015). *Social acceleration: A new theory of modernity* (J. Trejo-Mathys, Trans.). Columbia University Press.

Sakawa, M. (1993). *Fuzzy sets and interactive multiobjective optimization.* Plenum Press.

Sloterdijk, P. (2011). *Spheres. Volume I: Bubbles: Microspherology* (W. Hoban, Trans.). Semiotext(e). (Original work published 1998).

Swimme, B., & Berry, T. (1992). *The universe story: From the primordial flaring forth to the Ecozoic era—A celebration of the unfolding of the cosmos.* Harper San Francisco.

Index

L. de Miranda, *Crealectics as a Creative Method*,
Palgrave Advances in Possibility Studies,
https://doi.org/10.1007/978-3-032-00427-7

The manufacturer's authorised representative in the EU is Springer Nature Customer Service Centre GmbH, Europaplatz 3, 69115 Heidelberg, Germany. If you have any concerns regarding our products, please contact ProductSafety@springernature.com

Printed and bound by CPI Group (UK) Ltd, Croydon, CR0 4YY
10/06/2026
02134249-0001